CRYPTO SCIENTIFIC

MINING FOR DIGITAL GOLD

By

KALEI POTEAT

Lucid Definitions Publishing

Kalei Poteat is NOT a United States Securities Dealer or Broker, or U.S. Investment Adviser. Nor is he a sender a trader, dealer, or broker of ANY jurisdiction and or country. Sender makes no warranties or representations as to the Buyer, Seller, or any Transaction that may be contained or disclosed herein. The materials on Kalei Poteat website are provided "as shown." Kalei Poteat makes no warranties, expressed or implied, and hereby disclaims and negates all other warranties, including without limitation, implied warranties or conditions of merchantability, fitness for a particular purpose, or non-infringement of intellectual property or other violation of rights. Further, Kalei Poteat does not warrant or make any representations concerning the accuracy, likely results, or reliability of the use of the materials on its internet website or otherwise relating to such materials or on any sites linked to this site. RISK LEVELS Kalei Poteat ("Kalei Poteat" or "Kalei" or "we" or "us") is not registered as an investment adviser. Tom Gentile relies upon the "publishers' exclusion" from the definition of investment adviser under Section 202(a)(11) of the Investment Advisers Act of 1940 and corresponding state securities laws. As such, Kalei Poteat does not offer or provide personalized advice. We publish information about companies in which we believe our readers may be interested, and this information reflects our sincere opinions. The information that we provide or that is derived from our website is not intended to be, and should not be construed in any manner whatsoever as, personalized advice. Also, our website and the information provided by us should not be construed by any subscriber or prospective subscriber as Kalei Poteat's solicitation to effect, or attempt to effect, any transaction in a security. Investments in the securities markets, and especially in options and futures, are speculative and involve substantial risk. The information that we provide or that is derived from our website should not be a substitute for advice from an investment professional. We encourage you to obtain personal advice from your professional investment advisor and to make independent investigations before acting on the information that you obtain from Kalei Poteat or derive from our website/emails. Only you can determine what level of risk is appropriate for you. All trading involves risk. Leveraged trading has large potential rewards, but also large potential risk. Be aware and accept this risk before trading. Never trade with money you cannot afford to lose. All forecasting is based on statistics derived from past performance, and past performance of any trading methodology is no guarantee of future results. No "safe" trading system has ever been devised, and no one can guarantee profits or freedom from loss. Crypto Scientific: Mining For Digital Gold

No representation is being made that any account will achieve profits or losses similar to those discussed. There is no guarantee that, even with the best advice available, you will

TABLE OF CONTENTS

Chapter 1

Let's Talk $$$

What Is Money?

Money has single handedly placed itself as the staple of modern civilization. For decades, it has conceived its own language of propelling trade between nations, as a kind of language for value, and allowing mankind to store the product of their labor.

The definition of money is something that's widely accepted as payment for goods and services. I want you to think about what kind of goods and services you offer to your community and others. Ultimately, what do you do with your money to bring about change? Societies around the world have cultivated many varieties of money – so many, that it's difficult to neatly differentiate them all.

I would like to begin this book with a firm understanding about money and how we should think about the matter. Undoubtedly, the way we look at currency as a community has to change. Some of us have never been taught about using money to create wealth strategies, but instead have been indoctrinated into being lowly consumers. Let's observe the differences between commodity money, representative money, and fiat money.

Before money there was the Barter System

Barter refers to the act of exchanging goods or services for other goods or services. What are some goods and services that you spend your hard-earned money on that could possibility result in making you more money? When you spend, always remember that you need to be making an investment. Interestingly, this can also be observed in other contexts of life. Species in the plant and animal kingdoms enter unspoken agreements – *symbiotic* relationships – wherein both benefit from each other's actions. This is the science of mankind and the art of living together. For instance, Bullhorn acacia trees provide food and shelter for ants in exchange for protection from parasites. Wolves not only make carcasses available, but also open up tough hides that the birds would not be able to penetrate by themselves. Wolves often howl before going on hunt and ravens have learned to respond to this cue. It may be that wolves have also learned to listen to raven calls indicating that there are deer about. Symbiotic relationships can be useful measures of an ecosystem's health.

Of course, humans have a different and more sophisticated grasp of value than the above species. Long before the appearance of money as we currently know it, we understood that we could trade our goods for the goods of others.

It couldn't be more straightforward. Suppose that you have a blanket and your neighbor has oranges. She's freezing and you're hungry. You give her the blanket in exchange for twenty oranges. Both of you got what you wanted by trading it for what you have.

Painfully, it won't always be so simple. You'll probably want more oranges later, but your neighbor's new blanket will last her a couple of years. She may not want to trade again when you do. She might be out of luck herself if she wants to buy gas, but the gas station owner is allergic to oranges and won't take them.

The phenomenon is known as the *coincidence of wants* in economics. Bartering works well when you have something that your counterparty wants, and vice versa. It fails when parties don't need goods belonging to the other.

Commodities and Sustainability

Commodities are raw materials that are useful in some way (some might say they have *intrinsic value).* This definition encompasses an extensive range of things – from metals like gold, platinum, and tin to consumables like wheat, coffee, and rice.

It follows that commodity money involves the use of commodities as money. You're unlikely to be able to spend oil at your local store, but there are countless examples throughout history of useful raw materials being established as currency.

For instance, In the 17th and early 18th centuries, for example, American colonists used beaver pelts and dried corn in transactions. The further back in time that you can imagine, there was a form of commodity trading. This is one of the most primitive and ancient concepts of establishing value throughout the world.

On the surface, trading in commodities might not look too different from barter economies. After all, if you have a book such as this one and offer to sell it for beans, aren't you doing the same thing we discussed earlier?

Functionally, yes, but commodity money acts as a *medium of exchange.* In this scenario, you would expect beans to be widely accepted as a means of payment for goods or services. So unlike barter economies, where you trade goods or services for other goods or services, beans would be a tempting medium for exchange in many trades.

Commodities haven't gone away. As currencies, however, they've been largely replaced by other forms of money.

Represent Represent

Commodity money was a certain advancement to the barter system, but it still had shortcomings of its own, notably from a convenience perspective. While you could fit a handful of gold and diamonds in your pockets and use them to make small purchases, the concept doesn't scale.

Can you imagine trying to make medium-to-large purchases with nickels today? To give you an idea, if you wanted to buy a bitcoin from someone for 8,000 euros, you'd need to bring roughly 60 kg of one-euro coins with you.

After commodity money came representative money – a vastly more portable alternative backed by commodities. Representative money exploded in different places around the world at different times. In essence, it involves a central issuer producing certificates that can be redeemed for a certain amount of that commodity.

Rather than carry pounds of gold around, you could have pieces of paper that prove your ownership. You could go to the issuer and trade the paper in for actual gold at any moment. Or, you could give it to someone as payment and they could redeem it themselves. If you're familiar with stablecoins, this is basically the same principle.

The gold standard is a policy that was adopted by many governments where national currencies were backed by gold. Sure, there have been some instances of private firms issuing government-issued dollars, but it was done on a much larger scale by central banks. You might be familiar with the Gold Standard, which was a policy implemented by several countries that saw their national currencies backed by gold. While this sounds pretty outlandish in comparison to the system in place today, just under a century ago, you could bring your paper currency into the bank and exchange it for precious metals.

From an economic standpoint, this had some substantial benefits. So substantial that the gold standard is still a term used to describe something that outperforms alternatives. The first advantage was that, while government intervention increased, the currency wasn't easily debased through inflation. Governments couldn't issue banknotes exceeding the amount they were holding – in theory, at least. Unfortunately, it was easy (and attractive) for banks to operate a fractional reserve policy, where they created more notes than there was gold in storage.

The gold standard enabled individuals to transact with gold, without the inconvenience of carrying bullion around, or having to divide it up to make purchases.

Another benefit of this monetary system relates to the fact that gold has always been recognized on an international scale. If their domestic economies were gold-centric, then countries on the gold standard could more easily trade in a globally-appreciated resource.

FIAT MONEY

Representative money is no longer the dominant form of money. The death of the gold standard saw a new kind of money adopted globally – one totally detached from commodities. In short, fiat is money issued by a government (the term fiat stems from a Latin word meaning by *decree*). The United States dollars, Mexican pesos, Japanese yen, and Indian rupees used today are all examples of that. The value of fiat money is highly dependent on the decisions of governments and central banks. At its core, a fiat banknote is a piece of paper that only has value because governing bodies say so.

It should be noted that, while it's sometimes viewed as a recent invention, fiat in the form of paper money actually traces its origins back to 11th century China. It was also experimented with in 17th century Europe and the Americas for a couple of hundred years later. Unlike the types of money, we've discussed thus far, fiat currency has no element of scarcity. It's trivial to produce paper currency from readily-available materials, unlike agricultural commodities that take time to grow, or precious metals that must be extracted. Without these limitations, entities like the Federal Reserve can create new money out of thin air.

Depending on who you ask, this is either its greatest strength or its greatest weakness. Proponents of fiat money systems argue that the ability to inflate the money supply gives governments more flexibility for tackling financial crises or managing the overall economy. By controlling the money market and interest rates, the government has a high degree of control over the country's financial affairs.

Funnily enough, the same argument is put forward by opponents of fiat currency, albeit in a different light. One of the main criticisms of government monetary policy is that inflation slowly erodes the wealth of fiat money holders. Unchecked, it can bring on an aggressive period of inflation (hyperinflation), which essentially debases the currency entirely and can cause massive economic and social damage.

Crypto is the New Gold Standard

Bitcoin has been labeled as both digital cash and digital gold. On the one hand, it mimics many of the characteristics of money seen in commodities (i.e., fungibility, divisibility, portability), making it an ideal medium of exchange.

On the other, it is increasingly appreciated for store of value properties. Proponents of Bitcoin as digital gold believe that its supply policy of deflation (or more correctly, disinflation) will help it retain purchasing power over time. This stands in stark contrast to inflationary monies like the U.S. dollar, which can be devalued at the discretion of a Federal Reserve system.

Cryptocurrencies appear to fall under the category of commodity money. While they don't have any use outside of their protocols, they're not backed by anything or issued by government entities. The value of money, where digital currencies are concerned, stems from the valuation given to it by the free market.

So, what...?

Money takes forms of all types. Most of us are familiar with thinking about the value of our national fiat currencies, but we now have new technology and innovation on our hands. Fintech and apps like Robinhood have transitioned us thanks to years of economic evolution. We are able to trade crypto and stocks at the push of a button right from our smart devices, putting the powering into the hands of the common joe/jane.

Cryptocurrencies are a lucid conclusion to the next door of money. If Bitcoin or other Altcoins were to be considered and adopted, they will be the first real examples of digital commodities.

WHAT'S A CRYPTOCURRENCY?

A cryptocurrency is just like a digital form of cash. You can use it to pay friends for your share of a night on the town, buy that new pair of shoes you've been craving, or schedule the next travel booking for your honeymoon. Because cryptocurrency is digital, it can also be sent to friends and family anywhere in the world. Not to mention that it travels pretty swiftly.

You see, traditional online payment gateways are owned by organizations. They hold your money for you, and you need to ask them to transfer it on your behalf when you want to spend it. With cryptocurrencies, there isn't an organization. You, your friends, and thousands of others can act as your own banks by running free software. Your computer connects with other people's computers, meaning you communicate directly – no middlemen required! Perfect! To use cryptocurrency, you don't need to sign up for a website with an email address and password. You can download a wide variety of apps onto your smartphone to begin sending and receiving within minutes.

Why the name cryptocurrency?

The name **cryptocurrency** is a combo of ***cryptography*** and ***currency.*** With cryptography, we use advanced math to secure our funds, making sure that nobody else can spend them.

First thing we need to understand about this new technology is the fact that it is cheap and fast to use. Let's say you send money to the other side of the world. Your money can literally be there in the matter of second and at a fraction of the cost of what it would take to use a traditional international wire transfer. Also, based on the way that the network is designed, it's nearly impossible for hackers and security to be breached or shut down. Last, but not least, it is permissionless to the point that virtually no one can stop you from using cryptocurrency. Banks and centralized services, on the other hand, can hold funds by freezing your accounts or stop transactions from being made all together.

WHAT IS A CRYPTO WALLET?

The first thing you will need to have when it comes to crypto, is a wallet, a tool that you use to interact with a blockchain network. There are multiple types of wallets and they keep your crypto safe and sound. The three different groups of wallets are divided into software and hardware, also referred to as hot or cold wallets.

The majority of crypto wallet providers are based on software, which makes their use more convenient than hardware wallets. However, hardware wallets tend to be the most secure alternative. Paper wallets, on the other hand, consist of a "wallet" printed out on a piece of paper, but their use is now deemed as obsolete and unreliable.

As we think about how these wallet's work, crypto wallets don't truly store cryptocurrencies. Instead, they provide the tools required to interact with a blockchain. In other terms, wallets generate the information needed to send and receive cryptocurrency via blockchain transactions. This information consists of one or more pairs of public and private keys.

The wallet also includes an address, which is an alphanumeric identifier that is generated based on the public and private keys. Such an address is, in essence, a specific "location" on the blockchain to which coins can be sent to. This means you can share your address with others to receive funds, but you should never disclose your private key to anyone.

The private key gives access to your cryptocurrencies, regardless of which wallet you use. So even if your computer or smartphone gets compromised, you can still access your funds on another device – as long as you have the corresponding private key (or seed phrase). Note that the coins never truly leave the blockchain, they are just transferred from one address to another.

HOT VS. COLD WALLETS

As mentioned, cryptocurrency wallets may also be defined as "hot" or "cold," according to the way they operate.

A hot wallet is any wallet that connects to the Internet. When you create an account on Coinbase and proceed to send funds to it, then you are depositing into Coinbase's Hot Wallet. These types of wallets are easy to use and often time to go to for most investors. Cold wallets, on the other hand, have no connection with the net and is more likely a device that you can store the coins offline. This increased security for yourself by eliminating the risks associated with hackers. This is suitable for long term-investors. Some of the more popular Hot Wallets that are not concentrated on an exchange are Trust Wallet and Metamask. You can use web wallets to access blockchains through a browser interface without having to download or install anything. This includes both exchange wallets and other browser-based wallet providers.

In most cases, you can create a new wallet and set a personal password to access it. However, some service providers hold and manage the private keys on your behalf. Although this may be more convenient for inexperienced users, it's a dangerous practice. If you don't hold your private keys, you're trusting your money to someone else. To address this problem, many web wallets now allow you to manage their keys, either entirely or through shared control (via multi-signatures). So, it's important to check the technical approach of each wallet before choosing the most suitable for you.

Cold wallets, on the other hand, have no connection to the Internet. Instead, they use a physical medium to store the keys offline, making them resistant to online hacking attempts. As such, cold wallets tend to be a much safer alternative of "storing" your coins. This method is also known as cold storage and is particularly suitable for long-term investors or "HODLers." The most popular of these wallet types is Ledger X. You should consider using a hardware wallet if you plan to hold your crypto for a long time or if you're holding large amounts of cryptocurrency. Currently, most hardware wallets allow you to set up a PIN code to protect your device, as well as a recovery phrase – which can be used in case your wallet is lost.

When using cryptocurrency exchanges, you should consider making use of the protection tools available. The Coinbase Exchange offers several security features, such as device management, multi-factor authentication, anti-phishing code, and withdrawal address management. Losing access to your cryptocurrency wallets can be quite costly. So, it's important to back up them regularly. In many cases, this is achieved by simply backing up wallet.dat files or seed phrases. Essentially, a seed phrase works like a root key that generates and gives access to all keys and addresses in a crypto wallet. Also, if you opted for password encryption, remember to back up your password as well.

Crypto wallets are an integral part of using Bitcoin and other cryptocurrencies. They are one of the basic pieces of infrastructure that make it possible to send and receive funds through blockchain networks. Each wallet type has its advantages and disadvantages, so it's crucial to understand how they work before moving your funds.

Chapter 2

Unlocking the Block

Let's talk about Blockchains

When we talk about what the blockchain is, we have to know that it resembles a decentralized digital ledger as a list of data records. This data is organized into blocks, that are systematically arranged and bounded by encryption.

In the early 1990's, a computer Scientist by the name of Stuart Haber and a physicist named W. Scott Stonetter first utilized this strategy to secure digital documents from being tampered with. This in turn inspired many other computer wizards to study their work and eventually the birth of Bitcoin was cultivated.

Even though blockchain technology is older than cryptocurrencies, after Bitcoin sprung about in 2008, it's potential was being monitored. From then on, it's allure and greatness has been idolized on a global scale.

What makes the Blockchain tick?

Digging deeper into the blockchain's function, we must know that the blockchain is made up of a stable chain of blocks, each one storing a list of previously confirmed transactions. The network is maintained by the power of the people, linked across the planet. It acts as a databased stored on the computers of the world. Each participant also known as a node ("Bitcoin node" is a term used to describe a program that interacts with the Bitcoin network in some way), stores a copy of the blockchain data, letting each other know that each transaction is verified. This process is known as a block. There is no single authority in the matter and in term creates the decentralized system that we know and love.

When it comes to the mining process, we have to refer to the hashing algorithms. Bitcoin uses the SHA0256 algorithm. A secured hash algorithm of 256 bits. It takes an input of any length and generates an output that will always have the same length. This output is called a "hash" and in this case is always made of 64 characters (256bits). So, no matter how many times the process is repeated, the same input will push the same output. Changing the input is the only function that will create a new output, so in the crypto world, most of them are designed as a one-way hash function. This is adding security based on the fact that it is difficult to calculate what the input was to begin with.

Now that we know what the algorithm does, let's demonstrate how a blockchain works with a simple example of a transaction.

Imagine that we have Brianna and June along with their Bitcoin balance. Let's say Brianna owes June 2 Bitcoins. For Brianna to send June that 2 bitcoins, Brianna broadcasts a message with the transaction that she wants to make to all the miners in the network.

In that transaction, Brianna gives the miners June's address and the number of Bitcoins she would like to send, along with a digital signature and her public key. The signature is made with Brianna's private key and the miners can validate that Brianna, in fact, is the owner of those coins.

Once the miners are sure that the transaction is valid, they can put it in a block along with many other transactions and attempt to mine the block. This is done by putting the block through the SHA-256 algorithm. The output needs to start with a certain amount on 0's in order to be considered valid. The amount of 0's needed depends on what's called the "difficulty" which changes depending on how much computing power there is on the network.

In order to produce an output hash with the desired amount of 0's in the beginning, the miners add what's called a "nonce" into the block before running it through the algorithm. Since a small change to the input completely changes the output, the miners try random nonces until they find a valid output hash.

Once the block is mined the miner broadcasts that newly mined block to all the other miners. They then check to make sure that the block is valid so that they can add it to their copy of the blockchain and the transaction is complete. But in the block, the miners also need to include the output hash from the previous block so that all blocks are tied together, hence the name blockchain. This is an important part because of the way trust works in the system.

Every miner has their own copy of the blockchain on their computer and everyone trusts whichever blockchain that has the most computational work put into it, the longest blockchain. If a miner changes a transaction in a previous block, the output hash for that block will change which leads to all the hashes after it changing as well due to the blocks being liked with hashes. The miner would have to redo all of the work in order to make anyone accept his blockchain as the right one. So, if a miner wanted to cheat, he would need more than 50% of the networks computing power which is very unlikely. Network attacks like this are thereby called 51% attacks.

The model of making computers work in order to produce blocks is called Proof-of-Work (PoW) there are also other models like Proof-of-Stake (PoS) which does not require as much computing power and is meant to require less electricity while being able to scale to more users.

The Blockchains use advanced cryptographic techniques to remain secure through mathematical models of behavior and decision-making. This keeps digital money from being duplicated or destroyed.

Bitcoin's Creation

Amazingly, nobody knows who invented Bitcoin. We only know them by their screen name - **Satoshi Nakamoto**. Satoshi could be a single person, a group of programmers, or if you believe some of the weirder theories, a time-traveling alien or secret government team.

Bitcoin's creator is infamous and no one really knows who he is. All they know is that his name is Satoshi Nakamoto which could be a single entity or a group of programmers. There are many conspiracies surrounding this name. In 2008, a 9-page document was published detailing how the Bitcoin system worked and in 2009 the software was released.

There are many differences between Bitcoin and other cryptos such as speed, programmability, or privacy. Just continue to do research and your due diligence when entering this arena.

What is Bitcoin?

Bitcoin is a digital form of cash. But unlike the fiat currencies you're used to, there is no central bank controlling it. Instead, the financial system in Bitcoin is run by thousands of computers distributed around the world. Anyone can participate in the ecosystem by downloading open-source software.

Bitcoin was the first cryptocurrency, announced in 2008 (and launched in 2009). It provides users with the ability to send and receive digital money (bitcoins, with a lower-case *b*, or *BTC*). What makes it so attractive is that it can't be censored, funds can't be spent more than once, and transactions can be made at any time, from anywhere.

What is Bitcoin used for?

Bitcoin became popular from illicit activities such as fraud and the escorting side of things. When Backpage.Com stopped processing Visa Payments, Bitcoin became the main source of currency when advertising escorts. Also, the dark web and communities like Silk Road would accept Bitcoin as payment for a wide range of illegal documents, drugs, or malicious software.

People use Bitcoin for a number of other reasons today. Many appreciate it for its *permissionless* nature – anyone with an Internet connection can send and receive it. It's a bit like cash in that no one can stop you from using it, but its digital presence means that it can be transferred globally.

Bitcoin's value comes from its ability to be is decentralized, censorship-resistant, secure, and borderless. This quality has made it appealing for use cases such as international remittance and payments where individuals don't want to reveal their identities (as they would with a debit or credit card).

Many don't spend their bitcoins, instead choosing to hold them for the long-term (also known as *hodling*), striking a new slogan known as Diamond Hands. Bitcoin has been nicknamed *digital gold*, due to a finite supply of coins available. Some investors view Bitcoin as a perfect opportunity to inflate their portfolios. Because it's scarce and difficult to produce, it has been likened to precious metals like gold or silver.

Holders believe that these traits – combined with global availability and high liquidity – make it an ideal medium for storing wealth in for long periods. They believe that Bitcoin's value will continue to appreciate over time. Bitcoin to the moon!

Bitcoin at Work

When Alice makes a transaction to Bob, she's not sending funds in the way you'd expect. It's not like the digital equivalent of handing him a dollar bill. It's more like her writing on a sheet of paper (that everyone can see) that she's giving one dollar to Bob. When Bob goes to send those same funds to Carol, she can see that Bob has them by looking at the sheet.

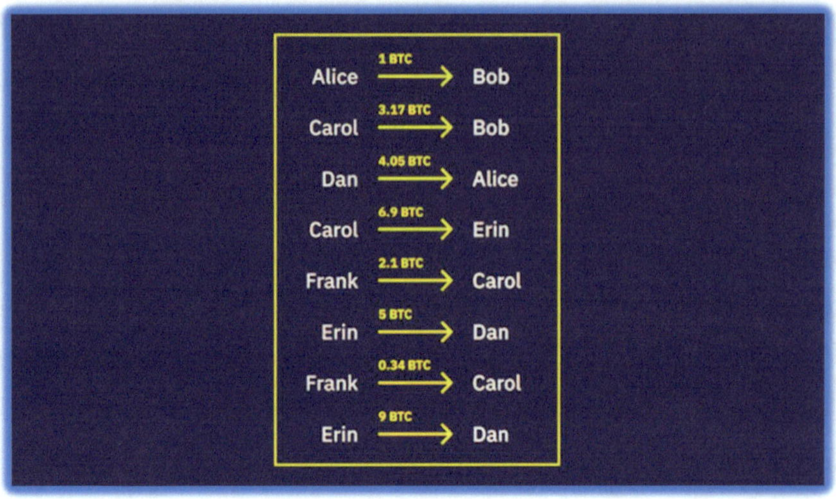

The sheet is a particular kind of database called a blockchain. Network participants all have an identical copy of this stored on their devices. The participants connect with each other to synchronize new information.

When a user makes a payment, they broadcast it directly to the peer-to-peer network – there isn't a centralized bank or institution to process transfers. In order to add new information, the Bitcoin blockchain uses a special mechanism called mining. It is through this process that new blocks of transactions are recorded in the blockchain.

Mining Bitcoins

Bitcoin has a finite supply, but not all units are in circulation yet. The only way to create new coins is through a process called *mining* – the special mechanism for adding data to the blockchain.

The protocol fixes Bitcoin's max supply at twenty-one million coins. As of 2020, just under 90% of these have been generated, but it will take over one-hundred years to produce the remaining ones. This is due to periodic events known as halving, which gradually reduce the mining reward

Bitcoin has a finite supply, but all have yet to be seen. In order to create new coins into circulation, there is a process called Mining. This is how data is added to the blockchain. In order to complete this task, computer power is used to solve cryptographic puzzles. For being dedicated to the cause, you can be rewards for introducing a valid block. This reward is made up of fees attached to the transaction. This protocol adjusts the intensity of mining so that it could take upward of ten minutes to find a new block. Mining Bitcoin today requires significant investment – not only in hardware but also in energy. At the time of writing, a good mining device performs upwards of ten trillion operations per second. Although very efficient, ASIC miners consume tremendous amounts of electricity. Unless you have access to several mining rigs and cheap electricity, you're unlikely to ever turn a profit with Bitcoin mining.

What if I lose my bitcoins?

Because there's no bank involved, you're responsible for keeping your coins secure. Some prefer to store them on exchanges, while others take custody with a variety of wallets. If you use a wallet, it's crucial that you write down your seed phrase so that you can restore it.

The bank is not insuring your bitcoins as of now. There are companies that are being formed to provide crypto insurance policies, the same way the industry emerged to help insurance marijuana. It is up to you to keep your bitcoins and other cryptos safe either on an exchange, a digital wallet, or hardware wallet such as a Ledger X. In order to access your coins on a wallet, you will be provided a Seed phrase. It is in your best interest not to share this key phrase with anyone else.

Bitcoin Cash

When politics come into play, this can push new projects to the front. Let's introduce Bitcoin Cash (BCH) which was created by a group of investors, miners, entrepreneurs, and developers who couldn't come to terms with Bitcoin's original developmental scale. Orchestrated in August 2017, Bitcoin Cash launched as another peer-to-peer system that focused on lower transaction fees and increased scalability. At the time, Bitcoin was suffering from long transaction times and growing fees, taking away from the plans of it being instant payment. All network nodes needed to reach an agreement, giving this altcoin a faster and cheaper payment system more suitable for daily use.

While there are stores and merchants that accept Bitcoin Cash payments, it doesn't seem to be a widespread practice yet. As of June 2021, Bitcoin.com Map flags thousands of stores accepting BCH, but a great portion of them does not currently mention or provide such a payment option, which suggests that the map is inaccurate or outdated.

Among the thousands of cryptocurrency projects that forked from Bitcoin, BCH is one that managed to remain relatively relevant. While it may not have gone on to experience the same heights and notoriety as Bitcoin, you can still find stores that accept BCH as payment, particularly because of its lower transaction fees and faster confirmation times.

Still, larger block sizes also bring concerns about network security and, as such, Bitcoin is still considered the most secure blockchain network. Also, Bitcoin continues to be the most popular cryptocurrency, meaning that BCH has lower market liquidity and adoption than BTC.

What is Ethereum?

Ethereum is a decentralized computing platform somewhat like a PC or Laptop, but it doesn't run on any one single device.

Simultaneously, it cranks up through thousands of machines around the world. Like Bitcoin and other cryptos, it allows for transfer of digital money. If Bitcoin is Digital Gold, then Ethereum is Digital Oil or Digital Electricity perhaps being that it is capable of doing a lot more. You have the ability to deploy your own code, and interact with apps created by other users. This has led to the study of a new field I like to call Tokenomics. All sorts of sophisticated programs can be launched on Ethereum.

Simply put, the main idea behind Ethereum is that developers can create and launch code which runs across a distributed network instead of existing on a centralized server. This means that, in theory, these applications can't be shut down or censored.

Applications cannot be shut down or censored based on the fact that developers can create and launch code that runs across a distributed network instead of existing on a centralized server. These units of currency here are called Ether with Ethereum being the protocol itself. Since it works in such a way, programs cannot be misused by external parties. They are uploaded to the Ethereum database and programmed in a way that makes sure that the code cannot be edited. Pretty savvy if you ask me.

Anyone anywhere can launch applications that cannot be taken offline. Ether stores value through these applications and can set conditions on how value is transferred. The programs that make up these applications are called smarts contracts and be set to operate without human intervention. According to Wikipedia, all of this super galactic innovation came from Vitalik Dmitriyevich Buterin, a Russian-Canadian programmer who launched Ethereum in 2014 at the young age of 19.

THE DIFFERENCES OF BITCOIN AND ETHEREUM

So, what's the difference between Bitcoin and Ethereum? Bitcoin is referred to as a first-generation blockchain that has increased security since the system isn't as complex as one may think. It's ability to interact with applications outside of transactions is constrained. Ethereum came on the scene as the first of the second-generation of blockchains. Like Bitcoin, it acts in the same manner, but gives developers more flexibility with trying new code and created Decentralized Applications or DApps.

HOW DOES ETHEREUM WORK?

So, let's back up a little bit and talk about Smart Contracts. These are triggers by transactions by users or other contracts. Let's say I send a transaction to a contract, every node on the network runs the contract's code and records the output. This where the Ethereum Virtual Machine (EVM) comes in by converting the smart contracts into instructions that the computer understands. Technically, these smart contracts are just code that are smart enough to execute itself under certain conditions and enforcing agreements between parties.

WHAT IS A SMART CONTRACT?

A *smart contract* is just code. The code is neither smart, nor is it a contract in the traditional sense. But we call it smart because it executes itself under certain conditions, and it could be regarded as a contract in that it enforces agreements between parties.

Computer scientist Nick Szabo can be credited with the idea, which he proposed in the late 1990s. He used the example of a vending machine to explain the concept, stating that it could be viewed as a precursor to the modern smart contract. In the case of a vending machine, there is a simple contract being executed. Users insert coins, and in return, the machine dispenses a product of their choosing.

A smart contract applies this kind of logic in a digital setting. You could specify something simple in the code like *return "Money Maker!" when two ether is sent to this contract.*

In Ethereum, the developer would code this so that it can later be read by the EVM. They then publish it by sending it to a special address that registers the contract. At that point, anyone can use it. And the contract can't be deleted, unless a condition is specified by the developer when writing it.

Now, the contract has an address. To interact with it, users just need to send 2 ETH to that address. This will trigger the contract's code – all the computers on the network will run it, see that the payment has been made to the contract, and record its output *("Money Maker!")*.

The above is perhaps one of the most basic examples of what can be done with Ethereum. More sophisticated applications that connect many contracts can – and have – been built. When it comes to explain how Ether is created, it holds the same principle by rewarding the users that mine.

So, how is Ether created???

We briefly touched on mining earlier. If you're familiar with Bitcoin, you'll know that the mining process is integral to securing and updating the blockchain. In Ethereum, the same principle holds: to reward the users that mine (which is costly), the protocol rewards them with ether.

A large part of Ethereum's appeal is the ability for users to create their own assets on-chain, which can be stored and transferred like ether. The rules governing them are set out in smart contracts, allowing developers to set specific parameters regarding their tokens. These can include how many to issue, how to issue them, whether they're divisible, whether each is fungible, and many others. The most prominent of the technical standards that allow the creation of tokens on Ethereum is called ERC-20 – and that's why the tokens are popularly known as ERC-20 tokens.

Token functionality provides innovators with a vast playground for experimenting with applications on the cutting edge of finance and technology. From issuing uniform tokens serving as in-app currency, to producing unique ones backed by physical assets, there's a great deal of design flexibility. It's entirely possible that some of the best use cases for easy and streamlined token creation aren't even known yet.

How to buy ETH with a credit/debit card

Coinbase allows you to perfectly buy ETH in your browser. To do so:

- Go to the Buy and Sell Cryptocurrency portal.
- Select the cryptocurrency you want to buy (ETH), and the currency you wish to pay with.
- Log in to Coinbase, or register if you don't have an account.
- Select your payment method.
- If prompted, insert your card details and complete the identity verification process.
- That's it! Your ETH will be credited to your Coinbase account.

Ethereum is used for building DApps and as a tradable token with ether being the fuel for its ecosystem. Like traditional currency, it can be used to purchase goods and serves.

CAN I MAKE MONEY WITH ETHEREUM?

Ethereum is a volatile asset: you can easily lose & make money with it. Crypto's aims to be held for a long time to stimulate a global, programmable universe. Opposingly you can trade it against other Altcoins. Both methods carry financial risk.

WHAT IS DECENTRALIZED FINANCE (DEFI)?

DeFi aims to decentralize financial applications on public blockchains by making them free to access by anyone on the Internet. A new global financial system, it offers peer-to-peer networks & Decentralized Applications (DApps). This lets users maintain ownership of their funds at all times to create a unregulated financial system, free from limitations.

Currently, most of the applications are hard to use, the framework for these ecosystems is difficult. Smart contracts and blockchains have the potential to introduce groundbreaking technology designed to overcome large-scale adoption barriers.

What Is Etherscan?

Etherscan is a blockchain explorer for the Ethereum network. Free to use, it's accessible with the URL, www.Etherscan.io It allows users to search through blocks, wallet addresses, smart contracts, transactions, blocks, and other on-chain data. With Etherscan, users understand how to interact with wallets, DApps, and the Blockchain. It asks for a transaction ID (TXID), wallet address, and contract address to fill search fields.

Chapter 3

Coins and Tokenomics

Let's Take a Look at some of My Favorite Coins

What Is Polygon (MATIC)?

The future of online communities will be dominated by a multitude of interconnected networks, each with their own set of unique properties, trust assumptions, performance, and security.

Polygon introduced the framework for creating scalable solutions compatible with Ethereum. The Ethereum scaling roadmap is supported by Polygon. In the Inter-Blockchain Communication Protocol (IBC), messages can travel between different blockchains with implemented protocols. Similar to Polygon, this concept is geared toward the Ethereum ecosystem. It was introduced as the Matic Network, but later rebranded to Polygon.

The framework supports two types of Ethereum-compatible networks: secure & stand-alone chains. Secured chains rely on the infrastructure of attached chain, meaning each link doesn't have to adopt its own security model. A rollup is an example of a secured chain. In contrast, stand-alone chains have their own form of security. Ultimately, secured chains offer a higher level of reinforced security, while stand-alone chains offer flexibility for specific needs. An example of a stand-along chain would be a side chain.

What Is Polkadot (DOT)?

Polkadot is the leader of the next-generation blockchain protocol. This network works by connecting multiple specialized chains in one universal network. Currently, individual blockchains cannot communicate with one another. Firms are still attempting to "bridge" the blockchain: This would allow chain A to work with chain B and vice versa. Connecting thousands of blockchains at the same time is a long-term software development goal that may soon be a reality.

Polkadot was introduced by the co-founder of Ethereum, Dr. Gavin. In this evolved ecosystem, each individual blockchain is called a parachain, a parallel blockchain The main Polkadot chain is called the Relay Chain. Polkadot integrated Relay Chain and parachains to easily exchange information at anytime.

Substrate is a framework for creating cryptocurrencies and decentralized systems. On substrate any developer, can draft a custom parachain. Once connected to the Polkadot network, parachains can interact with eachother. Network validators secure data across the different parachains & ensure transactions can be shared to improve scalability.

Communities associated with a parachain will be able to govern their private network to meet the needs of its participants.

THE DOT TOKEN EXPLAINED

Polkadot has its own token, known as DOT. Serving as the platform networks' token, DOT has several uses. Firstly, it grants holders governance rights of the entire Polkadot platform. This means owners have a voice in determining network fees, voting on overall network upgrades & the deployment or removal of parachains.

DOT also facilitates the network consensus via staking. Token holders are incentivized to play by the rules or be at risk of losing their stake.

Lastly, users purchase the Polygon token to use for bonding. This is needed when new parachains are added to the Polkadot ecosystem. During the bonding period, the DOT is locked. It's released once the bond duration has ended and the parachain is removed from the ecosystem.

What Is Ripple?

Previously known as OpenCoin, Ripple is a payment and exchange network (RippleNet) on a distributed ledger database (XRP Ledger). It's main focus is to connect banks, payment providers, and digital asset exchanges, creating faster global payments. It was first idealized in 2004 by Ryan Fugger.

The cryptocurrency XRP was introduced alongside the Ripple Consensus Ledger (RCL) which was later renamed to XRP Ledger (XRPL). XRPL is a system that stores accounting information & provides exchange services across currency pairs. Secured by participants, it allows real-time financial transactions through a consensus mechanism. The XRP Ledger is not based on a Proof of Work consensus algorithm, nor does it rely on mining to verify transactions.

The Ripple network serves as a currency exchange system that focuses on global payment solutions for banks and financial institutions.

RippleNet may be implemented to improve the traditional payment systems. xCurrent allows for cost-efficient real-time payments across financial institutions, xRapid uses XRP as a bridge borderless currency to provide on-demand liquidity pools, and xVia facilitates the integration and communication of all RippleNet participants.

What Is Cardano (ADA)?

Cardano is a blockchain based on peer-reviewed academic research. It's currently developed by a team of engineers, mathematicians & scientists. Caradno's key design principles are security, scalability, and interoperability. ADA, the currency of Cardano, is used to perform operations on the Cardano blockchain, much like the relationship between ether (ETH) and Ethereum.

What is the Cardano (ADA) roadmap?

The roadmap is composed of five main phases: Byron, Shelley, Goguen, Basho, and Voltaire.

1. Byron marked the launch of the network's functionality, transferring ADA.
2. Shelley developed the brand further towards decentralization & the opportunity to stake pools run by ADA holders.
3. Goguen attempted to deploy functional smart contracts blockchain platform.
4. The Basho era promoted the optimization of scalability and interoperability.

5. The Voltaire introduced a treasury system to address governance.

HOW DOES CARDANO (ADA) WORK?

Cardano, aims to solve the scalability problems of generation one (e.g., Bitcoin) and generation two (e.g., Ethereum). Previous blockchains suffer from bottlenecks that limit user activity, making them inefficient worldwide use. Cardano aims to improve the crpytospace in with advanced platform pillars.

- Create a customized own Proof of Stake (PoS) consensus called Ouroboros to reduces energy costs while providing security guarantees.
- Improve scalability with Hydra & provide an enhanced user experience as new nodes are added to the network.
- The hard fork combinator allows forking without interruptions to the blockchain.

Not yet operational, Cardano will allow for scalable smart contract functionality. The network aims to be the base layer for more intricate applications to be founded upon. However even with their roadmap, Cardano has yet to deliver.

What is the ADA token?

ADA is Cardano's crypto token. 57.6% of the ADA supply was distributed in an Initial Coin Offering (ICO), raising $62.2 million. It's used as a digital currency to make transactions on the Cardano network. ADA holders have a stake in Cardano, which is used in stake pools & generate staking rewards.

How to store Cardano (ADA)?

Daedalus is a desktop software wallet for storing ADA. A full node wallet, Cardano must be downloaded to verify each transaction for maximum user security. Opposingly, Light wallets, don't require downloads. ADA can be stored on a cold storage hardware wallet such as AdaLite. Cardano aims to provide blockchain infrastructure in the entire crypto ecosystem.

What Is Solana (SOL)?

Solana is a blockchain network focused on fast transactions and high responsiveness with a unique ordering method. Users pay transaction fees with smart contracts powered by SOL, the Solana's form of cryptocurrency.

Scalability is the biggest challenges for blockchain technologies, however with widespread growth comes limitations in transaction speed & confirmation times. Solana tackles these limitations without compromising security or decentralization.

Using a method known as Proof of History (PoH), the Solana blockchain handles thousands of transactions per second.

How does Solana work?

A Proof of Stake blockchain, it implements a trustless system for determining the time of a transaction, otherwise known as Proof of History.

Keeping track of transactions is vital for cryptocurrencies. For instance, Bitcoin bundles transactions into blocks with a single timestamp with each node being validated. Working in consensus with other nodes, the info blocks add a waiting time frame to confirm across a network.

What is Proof of History?

Solana transactions are linked using the SHA256 hash function. The function uses an input and produces the output to be used as the input for the next hash. This process creates a continuous unbroken chain of hashed transactions, providing a clear, verifiable order that a validator can then add to a block, without no conventional timestamp. Hashing requires a certain amount of time, making it easy for validators to verify how much time has passed. By viewing transactions in a chain of hashes, validators can process and transmit less information in each block.

Proof of History is not a consensus mechanism, rather a way to improve the time spent confirming a order of transactions. When combined a proof of stake, nodes require less time to validate & the network is able to choose a new validator quicker than previously.

SOLANA'S KEY FEATURES

The Solana team has developed eight core features to match the capabilities of a centralized system. Their technical application to the blockchain incorporates Proof of History in conjunction with:

- Tower BFT — A PoH version of Practical Byzantine Fault Tolerance
- Turbine — a block propagation protocol
- Gulf Stream —A Mempool-less transaction forwarding protocol
- Sealevel — A Parallel smart contracts run-time
- Pipelining — A Transaction Processing Unit for validation optimization
- Cloudbreak — A Horizontally-Scaled Accounts Database
- Archivers — A Distributed ledger storage

Collectively, these features create a high-performance network, operating thousands of transactions each second. This is favorable in comparison to the block time of Bitcoin (10 minutes) & Ethereum (15 seconds).

SOL holders can stake tokens as part of the PoS consensus mechanism. With a crypto wallet, users can stake tokens with validators who then process the network's transactions. Once successful, a validator can share rewards with users who staked. This, in turn, incentivizes delegators to act in the interest of Solana. In May 2021, Solana expanded to 900 validators, making it a fairly decentralized network.

What is SOL token?

SOL is Solana's cryptocurrency. A utility token, SOL pays transaction fees during transfers or with smart contracts. In this deflationary model, SOL holders can become network validators. Solana allows developers to create smart contracts and projects based on the blockchain using the SPL protocol. A token standard, SOL token has two main uses:

- Paying for incurred transaction fees.
- Staking tokens as part of the Proof of Stake consensus mechanism.

How to store SOL?

Holders can store SOL tokens on the sollet.io crypto wallet or any SPL-supporting wallet. To stake SOL, use a wallet that supports staking such as SolFlare. This will allow you to create a stake account and delegate SOL tokens to a validator.

Solana aims to provide speed and scalability benefits. Even with a well-priced token, network adoption and usage is still in its early stages.

What Is Uniswap and How Does It Work?

Uniswap is a computer program that run on the Ethereum blockchain that promotes decentralized token swaps. Traders can exchange Ethereum tokens on Uniswap sharing account information. Alternatively, users have the option to lend crypto in special reserves referred to as liquidity pools. In exchange for contributing money to into these pools, users earn interest income.

Centralized exchanges are the backbone of the cryptocurrency marke. They provide fast settlement times, high trading volume, and continually improving liquidity. In contrast, decentralized exchanges (DEX) require no middlemen to facilitate trading.

With the current limitations of blockchain technology, it's a challenge to build DEXes that compete with their centralized counterparts. Most DEXes hold room for improvement in terms of performance and user experience. Developers are constantly thinking of new ways to build a decentralized exchange: Uniswap is considered the pioneer DEX platform.

What is Uniswap?

Uniswap is the most successful project in the Decentralized Finance (DeFi) built on Ethereum. At its core, it is an **automated liquidity** protocol: No order book or party is required to make trades. Uniswap allows trades to occur with no intermediaries & a high degree of censorship-resistance.

Uniswap is open-source software. It works with liquidity providers that create liquidity pools. This system creates a decentralized pricing mechanism, seamlessly swapping between tokens without the need for an order book.On Uniswap there is no listing process. Any ERC-20 token can be launched if a liquidity pool is available for traders.

How does Uniswap work?

Uniswap has no order book & doesn't charge listing fees, it works with a design called Constant Product Market Maker, a variant of the Automated Market Maker (AMM). Automated market makers retain liquidity reserves that trader can trade against. Reserves are funded by liquidity providers that deposit at least two tokens in the pool. Traders then pay a pool fee to be distributed to liquidity providers according to individual shares.

How does Uniswap make money?

It doesn't. Uniswap is a decentralized protocol. Backed by Paradigm, all earned fees go to liquidity providers, not the platform creators.

Currently, the transaction fee paid out to liquidity providers is 0.3% per trade. By default, these are added to the liquidity pool, but liquidity providers can redeem them at any time. The fees are distributed according to each liquidity provider's share of the pool.

A portion of fees may be dedicated to Uniswap development in the future. The Uniswap team has already deployed an improved version of the protocol called Uniswap v2.

How to use Uniswap

Uniswap is an open-source protocol that anyone can create their own frontend application for. The most commonly used interface is https://app.uniswap.org or https://uniswap.exchange.

1. Go to the Uniswap interface.
2. Connect a wallet such as MetaMask, Trust Wallet, or any supported Ethereum wallet.
3. Select the token you'd like to exchange from.

4. Select the token you'd like to exchange to.

5. Click Swap.

6. Preview the transaction in the pop-up window.

7. Confirm the transaction request in your wallet.

8. Wait for the transaction to be confirmed on the Ethereum blockchain.

9. Monitor its status on https://etherscan.io/.

THE UNISWAP (UNI) TOKEN

UNI is the token of the Uniswap protocol & it entitles holders to governance rights. UNI holders can vote on changes to the internal protocol. The UNI token solidifies this idea of a public goods token. Currently, 1 billion UNI tokens have been minted & 60% of those are distributed to existing community members. The remaining 40% is available to team members, investors and advisors over the course of four years.

Community distribution occurs via liquidity mining. Essentially, UNI will be distributed to those who provide liquidity to the following Uniswap pools:

- ETH/USDT
- ETH/USDC
- ETH/DAI

- ETH/WBTC

Any Ethereum address that has interacted with a Uniswap contract is a community member.

HOW TO CLAIM UNISWAP (UNI) TOKENS

If you've previously used Uniswap, try to claim 400 UNI tokens per address that you've used Uniswap with. To claim Uniswap tokens:

- Go to https://app.uniswap.org/.
- Connect the wallet that you previously used Uniswap with.
- Click on "Claim your UNI tokens".
- Confirm the transaction in your wallet.

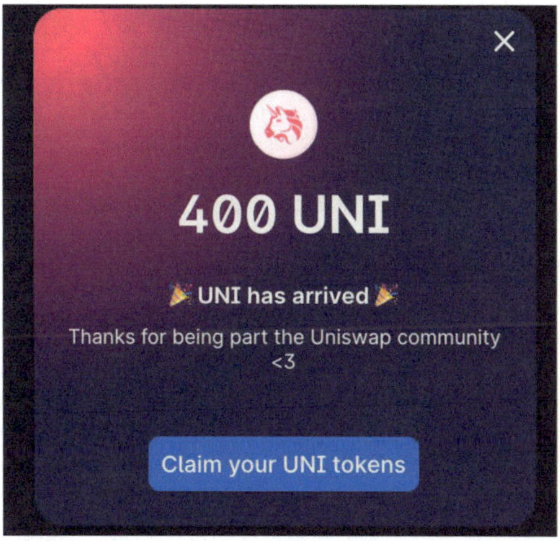

Uniswap is an innovative exchange protocol based on Ethereum. It allows anyone with an Ethereum wallet to exchange tokens without no 3rd party involvement.

It's implications have started the conversation of trustless token swapping. Uniswap is one of the core DeFi protocols.

WHAT IS SUSHISWAP?

Remember, Uniswap is a decentralized exchange protocol that works without an order book. SushiSwap is a version of Uniswap with key differences –notably, the SUSHI token. The SUSHI token has two integral functions: entitling holders to governance rights based on fees paid to the protocol. In essence SUSHI holders "own" the protocol.

Community governance is heavily intertwined with the DeFi network. Liquidity mining has created an abundance of new token launches. These tokens distributed grant governance rights to token holders. SUSHI holders are entitled to a portion of the fees paid into the protocol by traders.

With governance rights, anyone can submit a SushiSwap Improvement Proposal (SIP). Following a submission, SUSHI holders can then vote with their tokens. The development of SushiSwap lies in the hands of SUSHI token holders.

HOW ARE SUSHI REWARDS DISTRIBUTED?

SUSHI is distributed via liquidity mining to those who provide liquidity to specific Uniswap pools. Users can then deposit Uniswap LP tokens into SushiSwap staking contracts to start earning SUSHI.

Initially, Uniswap pools were:

- USDT-ETH
- USDC-ETH
- DAI-ETH
- SUSD-ETH
- COMP-ETH
- LEND-ETH
- SNX-ETH
- UMA-ETH
- LINK-ETH
- BAND-ETH
- AMPL-ETH
- YFI-ETH
- SUSHI-ETH

To incentivize the SUSHI market, the SUSHI-ETH pool pays out double rewards. Following its launch, SUSHI holders voted on adding additional pools to be eligible for distribution. Uniswap LP tokens staked on SushiSwap can be redeemed on Uniswap for the tokens they represent, and new liquidity pools will be created with them on SushiSwap.

HOW TO PROVIDE LIQUIDITY FOR SUSHISWAP

Getting the tokens

If you want to stake Uniswap tokens in exchange for SUSHI, the first step is to acquire UNI tokens. The following example provides liquidity for UMA-ETH, however any pair of tokens provided, that the LP tokens are usable on SushiSwap, is acceptable.

1. Head to Uniswap
2. Switch to the Pool tab
3. Click Add liquidity
4. The base should already be ETH, but if not, select ETH
5. Below, click on Select a token and search for UMA
6. Now enter the amount of ETH and UMA you are going to add to the liquidity pool

7. Finally, add the liquidity to the pool and you'll receive ETH-UMA UNI-V2 LP tokens – these can later be redeemed to retrieve your tokens once you have finished staking on SushiSwap.

Now that you have your **ETH-UMA UNI-V2 LP** tokens, visit sushiswap.org.

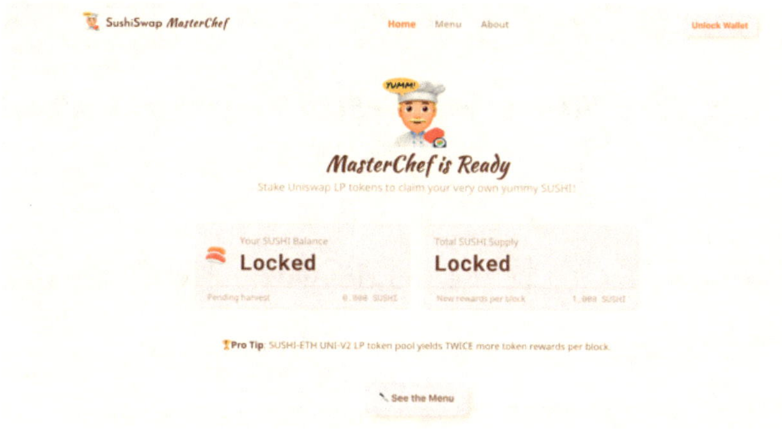

The man, the myth, the legend: Chef Nomi, the sushi chef.

First we must connect an Ethereum wallet to the site with either: MetaMask or WalletConnect. Select Unlock Wallet in the top-right corner, and pick your preferred wallet.

Unlocking the wallet shows you your SUSHI balance and the total supply.

Hit the Menu button at the top of the window to see what's available.

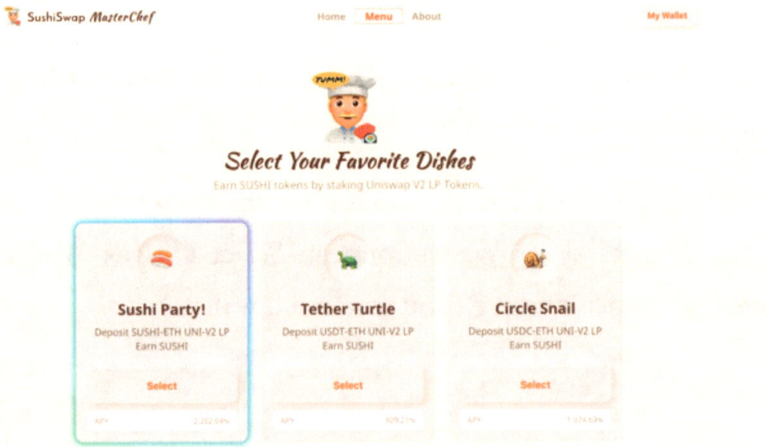

For each of these items, you'll stake a different token.

You may need to grant withdraw permissions to a contract. By selecting the approve function, we're telling the network that SushiSwap is allowed to take **UMA-ETH UNI-V2 LP** on our behalf.

After that, select the amount you want to stake.

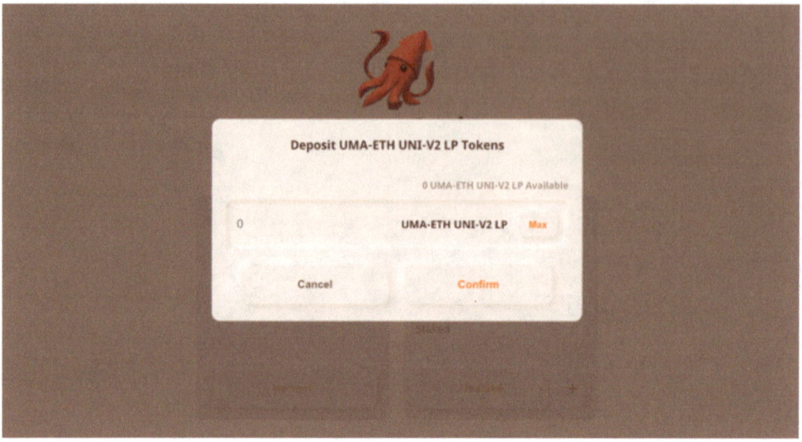

You're good to go! You can now harvest SUSHI, remove LP tokens, and take them back to Uniswap to retrieve your ETH and UMA.

IS SUSHISWAP SAFE?

Currently SushiSwap is an unaudited project. Depositing funds into a smart contract carries the risk of bugs for audited projects. SushiSwap was created by anonymous developers, which makes depositing funds into them even riskier. Never deposit more than you can comfortably lose. Smaller deposits have a lot of farming to do before they can actually turn a profit.

SushiSwap is an experiment thar challenges the competitive advantage of an the successful DeFi protocol – Uniswap. SushiSwap differs by providing community governance.

What Is Aave (AAVE)?

AAVE is the governance token of the Aave protocol. Holders can discuss and vote on proposals that affect the direction of the project. The AAVE token is one of the largest DeFi coins by market cap. Ethereum investors easily borrow and lend cryptocurrencies in a decentralized manner with Aave.

Modern financial ecosystem rely on mediums to borrow and lend assets. Borrowing allows users to leverage capital to accomplish tasks, while lending allows a safe return on an otherwise-idle capital. Cryptocurrency developers have acknowledged the need for financial services by launching money markets. Aave is one of the most successful of money marketplaces.

Aave is an Ethereum-based money market used to borrow and lend a variety of digital assets. The Aave protocol is governed by AAVE holders.

ETHLend

Aave's origins stretch back to 2017. Stani Kulechov and a team of developers released ETHLend in an Initial Coin Offering (ICO) in November 2017. The idea was to enable users to lend and borrow cryptocurrencies with each other by allowing users to post loan requests and offers.

While ETHLend was a novel idea, the platform, along with its token LEND, lost traction heading into the 2018 bear market. Key pain points with the platform were a lack of liquidity and the difficulty in matching loan requests to offers.

So, through the 2018 and 2019 bear market, the ETHLend team overhauled its product, releasing Aave at the start of 2020.

In a podcast, Kulechov said that the bear market was one of the best things that could have happened to ETHLend. This is in reference to the opportunity he and his team was given to revamp the concept of decentralized cryptocurrency lending, creating what we now know as Aave.

The new-and-improved Aave is similar in concept to ETHLend. Both allow Ethereum users to obtain cryptocurrency loans or earn a return by lending out their holdings. But, at their core, they are different.

Aave is an algorithmic money market, meaning loans are obtained from a pool instead of being individually matched to a lender.

The interest rate charged is dependent on the "utilization rate" of the assets in a pool. If nearly all assets in a pool are used, the interest rate is high to entice liquidity providers to deposit more capital. If nearly no assets in a pool are used, the interest rate charged is low to entice borrowing.

Aave also allows users to take out loans in a different cryptocurrency than they deposited. For instance, a user may deposit Ethereum (ETH), then withdraw stablecoins to deposit into Yearn.finance (YFI) to earn a regular yield.

Like ETHLend, all loans are overcollateralized. This means that if one wanted to borrow $100 worth of cryptocurrency through Aave, they would need to deposit more than that amount.

Due to the volatility of cryptocurrencies, Aave includes a liquidation process. If the collateral you provide falls under the collateralization ratio specified by the protocol, your collateral may be liquidated. Note that a fee is charged in case of liquidations. Make sure you understand the risks of depositing funds into Aave before posting collateral.

Aave is expanding its scope beyond just money markets. The platform has gained popularity as a place where DeFi users can obtain flash loans.

More often than not, there is much more liquidity in Aave's money-market pools than loans required by borrowers. This unused liquidity can be used by those that take flash loans, which are uncollateralized loans that only exist for the span of one Ethereum block.

Basically, a flash loan allows a user to borrow a large amount of cryptocurrency without posting collateral, then return the loan within the same transaction (as long as they pay the one-block interest fee).

This allows for those without large amounts of capital to arbitrage and enact other opportunities - all within a single blockchain transaction. For instance, if you see Ethereum trading for 500 USDC on Uniswap and 505 USDC on another decentralized exchange, you could try to arbitrage the price difference by borrowing a large amount of USDC and making quick trades.

Apart from flash loans and other features, Aave is also working on a non-fungible token (NFT) game called Aavegotchi.

What is Maker?

Maker (MKR) is the governance token of the MakerDAO and Maker Protocol — respectively a decentralized organization and a software platform, both based on the Ethereum blockchain — that allows users to issue and manage the DAI stablecoin.

Initially conceived in 2015 and fully launched in December 2017, Maker is a project whose task is to operate DAI, a community-managed decentralized cryptocurrency with a stable value soft-pegged to the US dollar.

MKR tokens act as a kind of voting share for the organization that manages DAI; while they do not pay dividends to their holders, they do give the holders voting rights over the development of

Maker Protocol and are expected to appreciate in value in accordance with the success of DAI itself.

The Maker ecosystem is one of the earliest projects on the decentralized finance (DeFi) scene: the industry that seeks to build decentralized financial products on top of smart-contract-enabled blockchains, such as Ethereum.

What Is Fetch.ai (FET)?

Fetch.ai is a platform that aims to connect Internet of Things (IoT) devices and algorithms to enable their collective learning. It was launched in 2017 by a team based in Cambridge, UK.

Fetch.ai is built on a high-throughput sharded ledger and offers smart contract capabilities to deploy machine learning and artificial intelligence solutions for decentralized problem-solving. Fetch.ai's utility token FET was designed to find, create, deploy and train autonomous economic agents and is an essential part of smart contracts and oracles on the platform.

Through the usage of FET, users can build and deploy their own agents on the network. Developers, by paying with FET tokens, can access machine-learning-based utilities to train autonomous agents and deploy collective intelligence on the network.

These open-source tools are designed to help users create ecosystem infrastructure and deploy commercial models.

What Is Near Protocol (NEAR)?

NEAR Protocol is a decentralized application platform designed to make apps usable on the web. The network runs on a Proof-of-Stake (PoS) consensus mechanism called Nightshade, which aims to offer scalability and stable fees.

NEAR is the native utility token that is used for:

- Fees for processing transactions and storing data.
- Running validator nodes on the network via staking NEAR tokens.
- Used for governance votes to determine how network resources are allocated.

NEAR tools include:

- NEAR SDKs which include standard data structures and testing tools for Rust and AssemblyScript.
- Gitpod for NEAR to create a zero-time onboarding experience for developers.
- NEAR Wallet that lets application developers create streamlined user experiences.
- NEAR Explorer to aid with both debugging of contracts and the understanding of network performance.

- NEAR Command Line Tools to allow developers to deploy applications from local environments.

What Is Chainlink (LINK)?

Founded in 2017, Chainlink is a blockchain abstraction layer that enables universally connected smart contracts. Through a decentralized oracle network, Chainlink allows blockchains to securely interact with external data feeds, events and payment methods, providing the critical off-chain information needed by complex smart contracts to become the dominant form of digital agreement.

The Chainlink Network is driven by a large open-source community of data providers, node operators, smart contract developers, researchers, security auditors and more. The company focuses on ensuring that decentralized participation is guaranteed for all node operators and users looking to contribute to the network.

What Is Dogecoin?

Dogecoin (DOGE) is based on the popular "doge" Internet meme and features a Shiba Inu on its logo. The open-source digital currency was created by Billy Markus from Portland, Oregon and Jackson Palmer from Sydney, Australia, and was forked from Litecoin in December 2013. Dogecoin's creators envisaged it as a fun, light-hearted cryptocurrency that would have greater appeal beyond the core Bitcoin audience, since it was based on a dog meme. Tesla CEO Elon Musk posted several tweets on social media that Dogecoin is his favorite coin.

Chapter 4

Stacking Your Coins

How to Build a Well-Balanced Crypto Portfolio

Balancing a crypto portfolio is not that different from balancing a traditional portfolio. You can easily reduce your overall risk according to your profile and investment strategy. All it takes to get started is simply diversifying your investments among different cryptocurrencies.

The extent to which you diversify is up for debate, as there are pros and cons to both sides. However, it's generally accepted that some diversification is beneficial. You can reduce the risk of your investments by holding different crypto assets (including stablecoins) and making sure to rebalance your asset allocation regularly.

To make managing your portfolio easier, you can use a third-party portfolio tracker or manually record your transactions on a spreadsheet. Some trackers can be linked to your personal wallets and cryptocurrency exchanges, making the process more convenient.

Getting started with crypto investing is as easy as purchasing your first bitcoin (BTC), ether (ETH), or any other cryptocurrency. While some investors prefer to buy and hold the largest cryptocurrencies, others choose to experiment with altcoins. But what's the best way to do this? You are more likely to succeed by thinking carefully about your asset allocation and balancing your crypto portfolio regularly. Depending on your risk tolerance, there are a few ways to do this. Balancing your portfolio isn't hard, and the results can literally pay off.

WHAT IS A CRYPTO PORTFOLIO?

A crypto portfolio is a collection of cryptocurrencies owned by an investor or trader. Portfolios typically contain a variety of different assets, including altcoins and crypto financial products. It's pretty similar to a traditional investment portfolio, except you're sticking to one asset class. You can track your crypto portfolio manually with a spreadsheet or use specialized tools and software to calculate your holdings and profits. A good portfolio tracker can come in handy. Trackers are essential to day traders and other short-term traders but also offer value to long-term investors and HODLers.

WHAT ARE ASSET ALLOCATION AND DIVERSIFICATION?

When creating an investment portfolio, you should be familiar with the concepts of asset allocation and diversification**.** **Asset allocation** refers to investing in different asset classes (e.g., cryptocurrencies, stocks, bonds, precious metals, cash, etc.). **Diversification** relates to the distribution of your investment funds across different assets or sectors. For example, you could diversify your stock holdings by investing in different industries, such as agriculture, technology, energy, and healthcare. Both of these strategies reduce your overall risk.

Technically, cryptocurrencies are a single asset class. But in a cryptocurrency portfolio, you can diversify across products, coins, and tokens that present different goals and use cases. For example, you could allocate your portfolio with 40% bitcoins, 30% stablecoins, 15% NFTs, and 15% altcoins. To explore the topic further, check out *Asset Allocation and Diversification Explained.*

CONCENTRATED VS. DIVERSIFIED CRYPTO PORTFOLIOS

Most advice will tell you that your crypto portfolio has to be diversified. While it is a standard for investors, there are pros and cons to spreading your capital around different assets. As we mentioned already, a diversified portfolio reduces overall risk and volatility. Losses can be offset by gains and keep your position stable. Your portfolio also has more opportunities to make gains with each coin you own. Not every investment will be a winner, but with proper asset allocation and diversification, you're more likely to make profits in the long run.

However, the more diversified your portfolio is, the closer it will track the overall market. Most traders and investors are looking to beat the market with larger gains. A highly diversified portfolio will lead to more average returns than a successful concentrated portfolio. Worse performing assets can balance out high earners.

Managing a diversified portfolio also requires more time and research. To invest soundly, you should understand what you are buying. With a large portfolio, the chances of understanding everything well decrease. If your portfolio is across different blockchains, you may also need to use multiple wallets and exchanges to access your assets. The decision to diversify or not is yours, but some diversification is always recommended.

DIFFERENT TYPES OF CRYPTOCURRENCIES

Bitcoin is the most famous cryptocurrency and is the biggest by market cap. But a well-balanced portfolio will include a selection of different coins to reduce overall risk. Let's go through some of them.

Payment coins

Nowadays, it's hard to find new coins that primarily deal in payments. But if you go back to the birth of cryptocurrencies, most projects were systems to transfer value. Bitcoin is the most well-known example, but we also have Ripple (XRP), Bitcoin Cash (BCH), and Litecoin (LTC), among others. These coins are the first generation of cryptocurrencies that existed before Ethereum and the introduction of smart contracts.

Stablecoins

A stablecoin attempts to track an underlying asset such as a fiat currency or precious metal. BUSD, for example, pegs the U.S. dollar with reserves set at a 1:1 ratio. PAX Gold (PAXG) uses the same system but ties the coin to the price of one fine troy ounce of gold held in reserves. While stablecoins don't necessarily provide large returns, they live up to their name and provide stability.

The cryptocurrency market is volatile, so having something in your portfolio that keeps its value is useful. If the stablecoin pegs something outside of the crypto ecosystem, a crypto market dip shouldn't affect it. If you want to move tokens out of a coin or project, you can rapidly transfer them to a dollar-backed stablecoin like BUSD to safeguard your gains. Converting into fiat is a much longer process than trading for a stablecoin.

Security tokens

Just like traditional securities, a security token can represent many things. It could be equity in a company, a bond issued by a project, or even voting rights. Securities have effectively been digitized and put on the blockchain, meaning that they mostly fall under the same regulations. For this reason, security tokens are in the jurisdiction of local regulators and must go through a legal process before issuance.

Utility tokens

A utility token acts as the key to a service or product. For example, BNB and ETH are both utility tokens. Among other things, you can use them to pay for transaction fees when interacting with decentralized applications (DApps). Many projects issue their own utility tokens to raise funds in a coin offering. The token's value should theoretically have a direct link to its utility's value.

Governance tokens

By holding a governance token, you can receive voting power on a project and even a share of the revenue. You'll most likely find these tokens in decentralized finance (DeFi) platforms like PancakeSwap, Uniswap or SushiSwap. Like utility tokens, the value of a governance token directly relates to the success of the underlying project.

FINANCIAL CRYPTO PRODUCTS

A portfolio doesn't just have to consist of holding different coins. Financial crypto products can also help diversify your portfolio even more. Think of it a bit like investing in government bonds, ETFs, or mutual funds rather than just holding shares. There's a massive number of products you can invest in across different blockchains and DApps.

Each investor or trader will have their own ideas on what makes a well-balanced crypto portfolio. But there are some general rules worth considering:

1. Split your portfolio between high, medium, and low-risk investments and give them appropriate weightings. A portfolio containing a large portion of high-risk investments is definitely not balanced. It might have the chance to provide you bigger gains but may also cause huge losses.

Your risk profile will determine what's best for you, but there should be some mix.

2. Consider holding some stablecoins to help provide liquidity for your portfolio. Stablecoins are the key to many DeFi platforms and can help you quickly and easily lock in gains or exit a position.

3. Rebalance your portfolio if needed. The crypto market is very volatile, and your decisions should change depending on the current situation.

4. Allocate new capital strategically to avoid overweighting any one area of your portfolio. If you've made big gains recently from one coin, it can be tempting to pump in more money. Don't let greed interfere, and think about where you can better place the money.

5. Do your own research. You really can't beat this classic piece of advice. You are investing your own money, so don't rely solely on the advice of others. For tips on spotting potential scams, see *5 Common Cryptocurrency Scams and How to Avoid Them.*

6. Only invest what you can afford to lose. Your portfolio isn't correctly balanced if you feel stressed about it. Your positions should not cause you serious consequences in case things go terribly wrong.

CRYPTO PORTFOLIO TRACKERS

A portfolio tracker is a program or service that allows you to trace the movements of your holdings. You can see how your current allocation stacks up with your long-term goals and track your progress. Here are some examples you might want to consider:

CoinMarketCap

CoinMarketCap is a hugely popular price tracker that's developed its own portfolio feature. The portfolio tracker is available for free on desktop and mobile devices. To use the portfolio tracker, you need to add in your holdings manually as it doesn't connect to your wallet or exchange. There's also the option to add in the prices you bought at to track your gains accurately.

CoinGecko

CoinGecko is mainly known for its cryptocurrency price tracking, but it has a portfolio option too. It's free to use and is available on your browser or mobile device. If you're already a regular CoinGecko user, the tracker is worth a try too.

Delta

Delta is a mobile app that allows you to view your crypto portfolio and traditional investments simultaneously. It can connect with 20 exchanges and a variety of wallets, including Binance. There's both a free and paid version, but you cannot trade within the app.

Chapter 5

Fail to Plan Then Plan to Fail

What Is a Bull Market?

Market trends are among the most fundamental aspects of financial markets. We can define a market trend as the overall direction that an asset or a market is going. As such, market trends are closely watched by both technical analysts and fundamental analysts.

Bull markets tend to be relatively straightforward to trade, as they can allow for some of the easiest trading and investment strategies. Even inexperienced traders may do well in really favorable bull market conditions. With that said, it's also crucial to understand how markets move in cycles.

A bull market (or bull run) is a state of a financial market where prices are rising. The term bull market is often used in the context of the stock market. However, it can be used in any financial market – including Forex, bonds, commodities, real estate, and cryptocurrencies. Besides, a bull market may also refer to a specific asset such as Bitcoin, Ethereum, or BNB. It could even refer to a sector, such as utility tokens, privacy coins, or biotech stocks.

Traders on Wall Street use the term "bullish" and "bearish" all the time. When they say that they are bullish on a market, it means that they think the price will rise. If they in turn say that they are bearish, they expect prices to drop. Being bullish can often mean that they want to hold a long position on that market. This doesn't mean that the long trade opportunity is there in the present, just that prices are expected to rise.

It's also worth noting that a bull market doesn't mean that prices don't fall or fluctuate. This is why it's more sensible to consider bull markets on larger time frames. In this sense, bull markets will contain periods of decline or consolidation without breaking the major market trend. Take a look at the Bitcoin chart below. While there are periods of decline, and a few violent market crashes, it has been in a major uptrend since its inception.

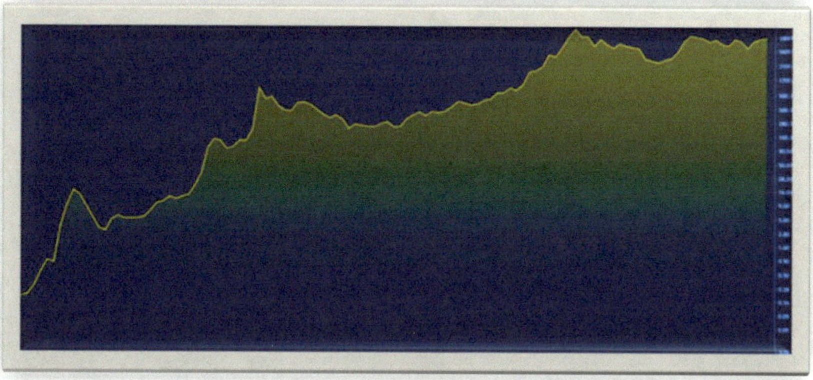

The Bitcoin price chart (2010-2020).

So, in this sense, the definition of a bull market depends on what time frame we're talking about. Generally, when we're using the term bull market, we are talking about a time frame of months or years. As with other market analysis techniques, higher time frame trends will have more validity than lower time frame trends.

Let's use an example of the long-term performance of the Dow Jones Industrial Average (DJIA). For a while, we can notice that is basically has been in a century-long bull market. There are instances where periods of decline can last for years, such as 1929 or 2008, but the overall trend is still pointing upwards.

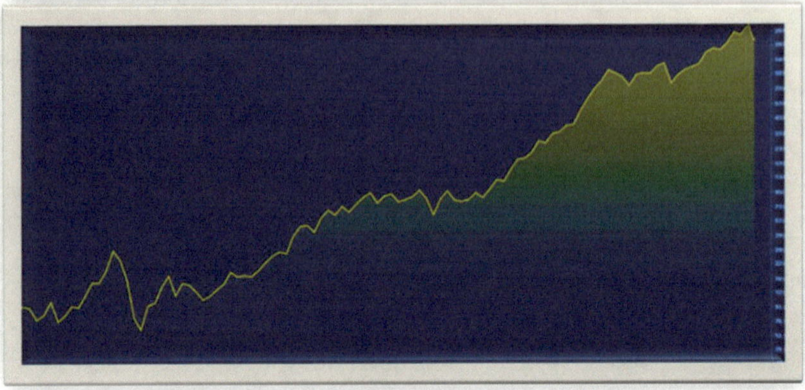

Performance of the DJIA since 1915.

What Is a Bear Market?

With financial markets moving in trends, it's important to understand the differences between these trends to be able to make better investment decisions. How come? Well, different market trends can lead to wildly different market conditions. If you don't know what the underlying trend is, how are you going to adapt to changing conditions?

A bear market can be described as a period of declining prices in a financial market. Bear markets can be extremely risky and difficult to trade for inexperienced traders. They can easily lead to great losses and scare investors from ever returning to the financial markets. How come?

There's this saying among traders: "Stairs up, elevators down." This means that moves to the upside may be slow and steady, while moves to the downside tend to be more sharp and violent. Why is that? When the price starts crashing, many traders rush to exit the markets. They do that to either stay in cash or lock in profits from their long positions. This can quickly result in a domino effect where sellers rushing to the exit leads to even more sellers exiting their positions, and so on. The drop can be amplified even more if the market is highly leveraged. Mass liquidations will have an even more pronounced cascading effect, resulting in a violent sell-off.

Typically, investors are "bearish" in a bear market, meaning that they expect prices to decline. This also means that market sentiment is generally quite low. However, this may not mean that all market participants are in active short positions. This just means that they expect prices to decline and may be looking to position themselves accordingly if the opportunity presents itself.

After Bitcoin's move to around $20,000 in December 2017, it's had quite a brutal bear market.

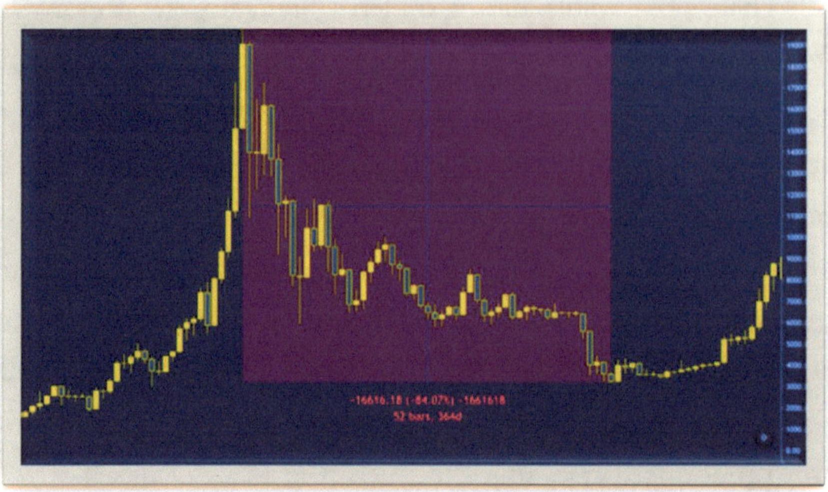

Bitcoin price crashes after the 2017 bull market.

And before the 2018 bear market, Bitcoin experienced an 86% drop in 2014.

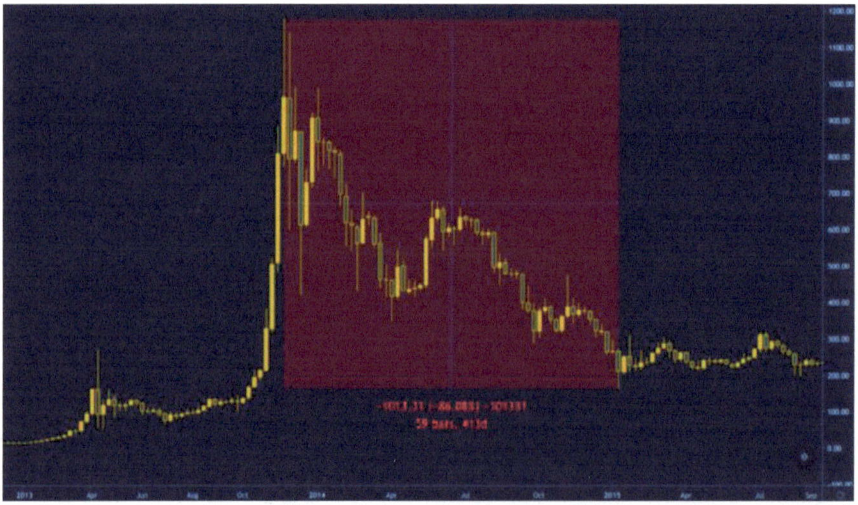

Bitcoin price crashes 86% from the 2013 top.

As of July 2020, the range of the previous bear market low around $3,000 have been retested but never broken. If that low would have been breached, a stronger argument could be made that a multi-year Bitcoin bear market is still underway.

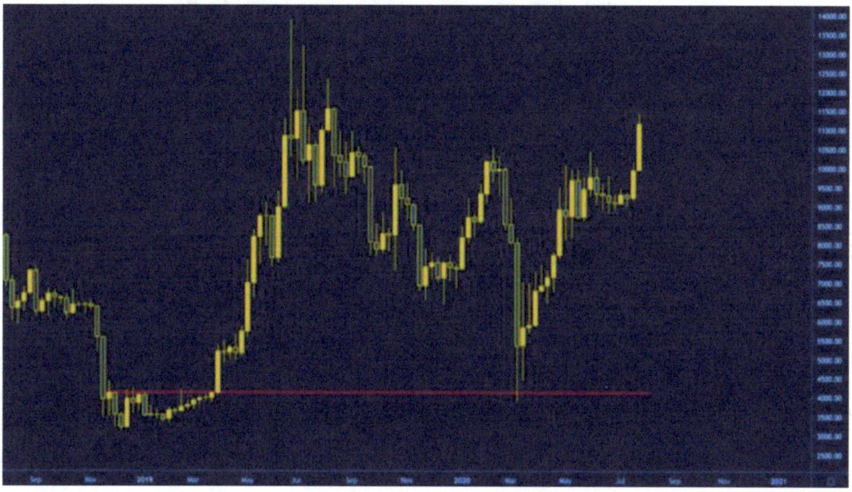

Bitcoin retesting the range of its previous bear market low.

Since that level has not been broken, the argument can be made that the crash following COVID-19 fears was merely a retest of the range. Still, there are no certainties when it comes to technical analysis, only probabilities.

Other notable bear market examples come from the stock market. The Great Depression, the 2008 Financial Crisis, or the 2020 stock market crash due to the coronavirus pandemic are all noteworthy examples. These events have all caused great damage on Wall Street and impacted stock prices across the board. Market indexes such as the Nasdaq 100, the Dow Jones Industrial Average (DJIA), or the S&P 500 index can experience significant price declines during times like these.

What Is a Short Squeeze?

Traders can profit from a price decline of an asset by short selling. It's a very common way to manage downside risk, hedge existing holdings, or simply express a bearish outlook on the market. Shorting an asset is a highly risky trading strategy when you are taking a position that the asset's price will go down. There is no upper limit for the price of an asset, but you also have to deal with what is called a short squeeze. When there is a sudden price increase a short squeeze can occur and many short sellers get trapped. In order for them to cover their positions, they have to quickly exit. Many traders have stop-loss triggers on or they may opt for closing their positions themselves so that they won't lose any more money. So, when you are shorting an asset, you sell it first and buy last. As you begin to understand the market you are in, you can begin to identify short squeezes and go long on a position for profit. Short squeezing happens in the stock market and also with the Bitcoin market all the time.

Take a look at this Bitcoin price range below from early 2019. The price was contained in a range after a sharp move to the downside. Market sentiment was likely quite low, as many investors would be looking for short positions, expecting the continuation of the downtrend. Many traders took short positions on Bitcoin in 2019, hoping that the price would continue to fall. The huge move up most likely came from people trying to cover their short positions.

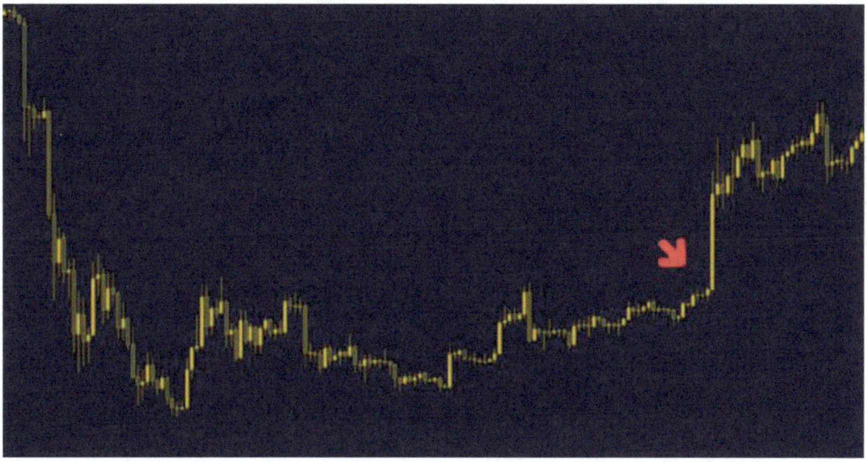

Potential short squeeze on the BTC/USD market.

TRADING

When it comes to trading assets, you look at the short-term aspect of getting into a position on an underlying asset and getting out rather quickly to make some money. These assets are also known as financial instruments. These can be bonds, currency pairs on the Forex market, stocks, options, futures, cryptocurrency, and more. The key is to know when to get in and when to get out. Instead of speculating when is a good entry and exit point, the use of Technical

Analysis (TA) and Fundamental Analysis (FA) is highly recommended. Technical analysts look at price history, charts, and other types of market data to find bets that have a good chance of returning a profit.

What is Investing?

Investing is allocating resources (such as capital) with the expectation of generating a profit. This can include using money to fund and kickstart a business or buying land with the goal of reselling it later at a higher price. In the financial markets, this typically involves investing in financial instruments with the hopes of selling them later at a higher price.

Trading and Investing are two separate ideas. Investing is allowing yourself to allocate you capital or money to potential generate profit on a longer time frame (years or even longer). Investing includes starting a business or buying property with the intentions of reselling at a high price.

The expectation of a return is core to the concept of investment (this is also known as ROI). As opposed to trading, investing typically takes a longer-term approach to wealth accrual. The goal of an investor is to build wealth over a long period of time (years, or even decades). There are plenty of ways to do that, but investors will typically use fundamental factors to find potentially

Trading vs. investing – what's the difference?

Both traders and investors seek to generate profits in the financial markets. Their methods to achieve this goal, however, are quite different.

Generally, investors seek to generate a return over a longer period of time – think years or even decades. Since investors have a larger time horizon, their targeted returns for each investment tend to be larger as well.

Traders, on the other hand, try to take advantage of the market volatility. They enter and exit positions more frequently, and may seek smaller returns with each trade (since they're often entering multiple trades).

Which one is better? Which one is more suitable for you? That's for you to decide. You can start educating yourself about the markets, and then learn by doing. Over time, you'll be able to determine which one suits better your financial goals, personality, and trading profile.

What is a trading journal?

A trading journal consists of a document where everything you do as a trader is recorded, including strategy development, risk management, psychology, and more. Keeping a trading journal is easy but very effective if created and used correctly. While it can bring valuable insights that may prevent your account from blowing up, it can also be the reason for your account to go to the moon.

There are several important reasons why keeping a trading journal is important, including:

- It keeps you accountable.
- It makes you more disciplined and consistent.
- It helps you spot profitable trading strategies.
- It documents your strengths and weaknesses.
- It makes you more careful when analyzing potential trades.

Successful traders plan all of their trades meticulously and document the successes and failures of their trading performance. By creating a trading journal and using it correctly, you can become a successful trader regardless of how the market moves.

How to Create a Trading Journal

You can find a free trading journal template in the next section, but it's also important to learn how to create one. You can customize your trading journal in a variety of formats to fit your trading style and needs. As long as you have somewhere to plan and document your trading activities, you'll be set.

First, you need to create a trading journal spreadsheet (e.g., Google Sheets, Microsoft Excel) and a written document (e.g., Google Docs, Microsoft Word). You'll be using these to record your exact trades and your thoughts, respectively. If you prefer, you can include the written document as a second tab in the spreadsheet (see template below).

Second, you'll need to know what you'll be recording daily so your trading journal will have the highest possible impact on your success. You can find several trading journals examples online. But regardless of the template, your spreadsheet should have columns related to each trade. These columns may include:

- Entry date
- Exit date
- Symbol
- Direction (long/short)
- Entry price
- Position size
- Notional value
- Stop loss
- Take profit
- Exit price

- Trading fees
- Profit/Loss (P&L)
- Profit/Loss percentage (P&L %)
- Notes

Some traders may also add the time frame, screenshots of the setup, and anything else they may deem important. The bottom line is for the information to work in their favor.

In your written document (or in another tab), you should have a section for each day where you can write down all your thoughts and ideas so you can get them organized.

The written document is where a trader lets loose their creativity while their spreadsheet helps to measure the profitability of their creativity. Both are very useful when creating and using a trading journal.

And that's it! Learning how to create a trading journal is the easy part, though. Knowing how to use a trading journal is something you'll need to get good at over time. However, as long as you have a strong understanding of the fundamentals, you'll be using your trading journal like a pro in no time.

Trading and Investment Strategies

What is a Trading Strategy?

A trading strategy is simply a plan you follow when executing trades. There's no single correct approach to trading, so each strategy will largely depend on the trader's profile and preferences.

Regardless of your approach to trading, establishing a plan is crucial – it outlines clear goals and can prevent you from going off course due to emotion. Typically, you'll want to decide *what* you're trading, how you're going to trade it, and the points at which you'll enter and exit.

In the following chapter, we'll get into a few examples of popular trading strategies.

What is Portfolio Management?

Portfolio management concerns itself with the creation and handling of a collection of investments. The portfolio itself is a grouping of assets – it could contain anything from Beanie Babies to real estate. If you're exclusively trading cryptocurrencies, then it will probably be made up of some combination of Bitcoin and other digital coins and tokens.

Your first step is to consider your expectations for the portfolio. Are you looking for a basket of investments that will remain relatively protected from volatility, or something riskier that might bring higher returns in the short term?

Putting some thought into how you want to manage your portfolio is highly beneficial. Some might prefer a *passive* strategy – one where you leave your investments alone after you set them up. Others could take an *active* approach, where they continuously buy and sell assets to make profits.

What is Risk Management?

Managing risk is vital to success in trading. This begins with the identification of the types of risk you may encounter:

- **Market risk:** the potential losses you could experience if the asset loses value.
- **Liquidity risk:** the potential losses arising from illiquid markets, where you cannot easily find buyers for your assets.
- **Operational risk:** the potential losses that stem from operational failures. These may be due to human error, hardware/software failure, or intentional fraudulent conduct by employees.
- **Systemic risk:** the potential losses caused by the failure of players in the industry you operate in, which impacts all

businesses in that sector. As was the case in 2008, the collapse of the Lehman Brothers had a cascading effect on worldwide financial systems.

As you can see, risk identification begins with the assets in your portfolio, but it should take into account both internal and external factors to be effective. Next, you'll want to assess these risks. How often are you likely to encounter them? How severe are they?

By weighing up the risks and figuring out their possible impact on your portfolio, you can rank them and develop appropriate strategies and responses. Systemic risk, for example, can be mitigated with diversification into different investments, and market risk can be lessened with the use of stop-losses.

What is Day Trading?

Day trading is a strategy that involves entering and exiting positions within the same day. The term comes from legacy markets, referencing the fact that they're only open for set periods during the day. Outside of those periods, day traders are not expected to keep any of their positions open.

Cryptocurrency markets, as you probably know, are not subject to opening or closing times. You can trade around the clock every day of the year. Still, *day trading* in the context of cryptocurrency tends to refer to a trading style where the trader enters and exits positions within 24 hours.

In day trading, you'll often rely on technical analysis to determine which assets to trade. Because profits in such a short period can be minimal, you may opt to trade across a wide range of assets to try and maximize your returns. That said, some might exclusively trade the same pair for years.

This style is obviously a very active trading strategy. It can be highly profitable, but it carries with it a significant amount of risk. As such, day trading is generally better suited to experienced traders.

What is Swing Trading?

In swing trading, you're still trying to profit off market trends, but the time horizon is longer – positions are typically held anywhere from a couple of days to a couple of months.

Often, your goal will be to identify an asset that looks undervalued and is likely to increase in value. You would purchase this asset, then sell it when the price rises to generate a profit. Or you can try to find overvalued assets that are likely to decrease in value. Then, you could sell some of them at a high price, hoping to buy them back for a lower price.

As with day trading, many swing traders use technical analysis. However, because their strategy plays out across a longer period, fundamental analysis may also be a valuable tool.

Swing trading tends to be a more beginner-friendly strategy. Mainly because it doesn't come with the stress of fast-paced day trading. Where the latter is characterized by rapid decision-making and a lot of screen time, swing trading allows you to take your time.

What is Position Trading?

Position (or trend) trading is a long-term strategy. Traders purchase assets to hold for extended periods (generally measured in months). Their goal is to make a profit by selling those assets at a higher price in the future.

What distinguishes position trades from long-term swing trades is the rationale behind placing the trade. Position traders are concerned with trends that can be observed over extended periods – they'll try to profit from the overall market direction. Swing traders, on the other hand, typically seek to predict "swings" in the market that don't necessarily correlate with the broader trend.

It's not uncommon to see position traders favor fundamental analysis, purely because their time preference allows them to watch fundamental events materialize. That's not to say technical analysis isn't used. While position traders work on the assumption that the trend will continue, the use of technical indicators can alert them to the possibility of a trend reversal.

Like swing trading, position trading is an ideal strategy for beginners. Once again, the long-time horizon gives them ample opportunity to deliberate on their decisions.

What is scalping?

Of all of the strategies discussed, scalping takes place across the smallest time frames. Scalpers attempt to game small fluctuations in price, often entering and exiting positions within minutes (or even seconds). In most cases, they'll use technical analysis to try and predict price movements and exploit bid-ask spread and other inefficiencies to make a profit. Due to the short time frames, scalping trades often give a small percentage of profits – usually lower than 1%. But scalping is a numbers game, so repeated small profits can add up over time.

Scalping is by no means a beginner's strategy. An in-depth understanding of the markets, the platforms you're trading on, and technical analysis are vital to success. That said, for traders that know what they're doing, identifying the right patterns and taking advantage of short-term fluctuations can be highly profitable.

What is asset allocation and diversification?

Asset allocation and diversification are terms that tend to be used interchangeably. You might know the principles from the saying *don't keep all your eggs in one basket*. Keeping all of your eggs in one basket creates a central point of failure – the same holds true for your wealth. Investing your life savings into one asset exposes you to the same kind of risk. If the asset in question was the stock of a particular company and that company then imploded, you'd lose your money in one swift movement.

This isn't just true of single assets, but of asset classes. In the case of a financial crisis, you'd expect all of the stock you hold to lose value. This is because they're heavily correlated, meaning that all tend to follow the same trend.

Good diversification isn't simply filling your portfolio with hundreds of different digital currencies. Consider an event where the world governments ban cryptocurrencies, or quantum computers break the public-key cryptography schemes we use in them. Either of these occurrences would have a profound impact on *all* digital assets. Like stocks, they make up a single asset class.

Ideally, you want to spread your wealth across multiple classes. That way, if one is performing poorly, it has no knock-on effect on the rest of your portfolio. Nobel Prize winner Harry Markowitz introduced this idea with the Modern Portfolio Theory (MPT). In essence, the theory makes the case for reducing the volatility and risk associated with investments in a portfolio by combining uncorrelated assets.

What is buy and hold?

The "buy and hold" strategy, perhaps unsurprisingly, involves buying and holding an asset. It's a long-term passive play where investors purchase the asset and then leave it alone, regardless of market conditions. A good example of this in the crypto space is *HODLing*, which typically refers to investors that prefer to buy and hold for years instead of actively trading.

This can be an advantageous approach for those that prefer "hands-off" investing as they don't need to worry about short-term fluctuations or capital gains taxes. On the other hand, it requires patience on the investor's part and assumes that the asset won't end up totally worthless.

What is index investing?

Index investing could be regarded as a form of "buy and hold." As the name implies, the investor seeks to profit from the movement of assets within a specific index. They could do so by purchasing the assets on their own, or by investing in an index fund.

Again, this is a passive strategy. Individuals can also benefit from diversification across multiple assets, without the stress of active trading.

What is paper trading?

Paper trading could be any kind of strategy – but the trader is only pretending to buy and sell assets. This is something you might consider as a beginner (or even as an experienced trader) to test your skills without putting your money at stake.

You may think, for instance, that you've discovered a good strategy for timing Bitcoin dips, and want to try profiting from those drops before they occur. But before you risk all of your funds, you might opt to paper trade. This can be as simple as writing down the price at the time you "open" your short, and again when you close it. You could equally use some kind of simulator that mimics popular trading interfaces.

The main benefit of paper trading is that you can test out strategies without losing your money if things go wrong. You can get an idea of how your moves would have performed with zero risk. Of course, you need to be aware that paper trading only gives you a limited understanding of a real environment. It's hard to replicate the real emotions you experience when your money is involved. Paper trading without a real-life simulator may also give you a false sense of associated costs and fees, unless you factor them in for specific platforms. WeBull offers a couple of options for paper trading.

Chapter 6

Let's Get Technical

What is a long position?

A long position (or simply long) means buying an asset with the expectation that its value will rise. Long positions are often used in the context of derivatives products or Forex, but they apply to basically any asset class or market type. Buying an asset on the spot market in the hopes that its price will increase also constitutes a long position.

Going long on a financial product is the most common way of investing, especially for those just starting out. Long-term trading strategies like buy and hold are based on the assumption that the underlying asset will increase in value. In this sense, buy and hold is simply going long for an extended period of time.

However, being long doesn't necessarily mean that the trader expects to gain from an upward movement in price. Take leveraged tokens, for example. BTCDOWN is inversely correlated to the price of Bitcoin. If the price of Bitcoin goes up, the price of BTCDOWN goes down. If the price of Bitcoin goes down, the price of BTCDOWN goes up. In this sense, entering a long position in BTCDOWN equals a downward movement in the price of Bitcoin.

What is shorting?

A short position (or short) means selling an asset with the intention of rebuying it later at a lower price. Shorting is closely related to margin trading, as it may happen with borrowed assets. However, it's also widely used in the derivatives market, and can be done with a simple spot position. So, how does shorting work?

When it comes to shorting on the spot markets, it's quite simple. Let's say you already have Bitcoin and you expect the price to go down. You sell your BTC for USD, as you plan to rebuy it later at a lower price. In this case, you're essentially entering a short position on Bitcoin since you're selling high to rebuy lower. Easy enough. But what about shorting with borrowed funds? Let's see how that works.

You borrow an asset that you think will decrease in value – for example, a stock or a cryptocurrency. You immediately sell it. If the trade goes your way and the asset price decreases, you buy back the same amount of the asset that you've borrowed. You repay the assets that you've borrowed (along with interest) and profit from the difference between the price you initially sold and the price you rebought.

So, what does shorting Bitcoin look like with borrowed funds? Let's look at an example. We put up the required collateral to borrow 1 BTC, then immediately sell it for $10,000. Now we've got $10,000. Let's say the price goes down to $8,000. We buy 1 BTC and repay our debt of 1 BTC along with interest. Since we initially sold Bitcoin for $10,000 and now rebought at $8,000, our profit is $2,000 (minus the interest payment and trading fees).

What is the order book?

The order book is a collection of the currently open orders for an asset, organized by price. When you post an order that isn't filled immediately, it gets added to the order book. It will sit there until it gets filled by another order or canceled.

Order books will differ with each platform, but generally, they'll contain roughly the same information. You'll see the number of orders at specific price levels.

When it comes to crypto exchanges and online trading, orders in the order book are matched by a system called the matching engine. This system is what ensures that trades are executed – you could think of it as the brain of the exchange. This system, along with the order book, is core to the concept of electronic exchange.

What is the order book depth?

The order book depth (or market depth) refers to a visualization of the currently open orders in the order book. It usually puts buy orders on one side, and sell orders on the other and displays them cumulatively on a chart.

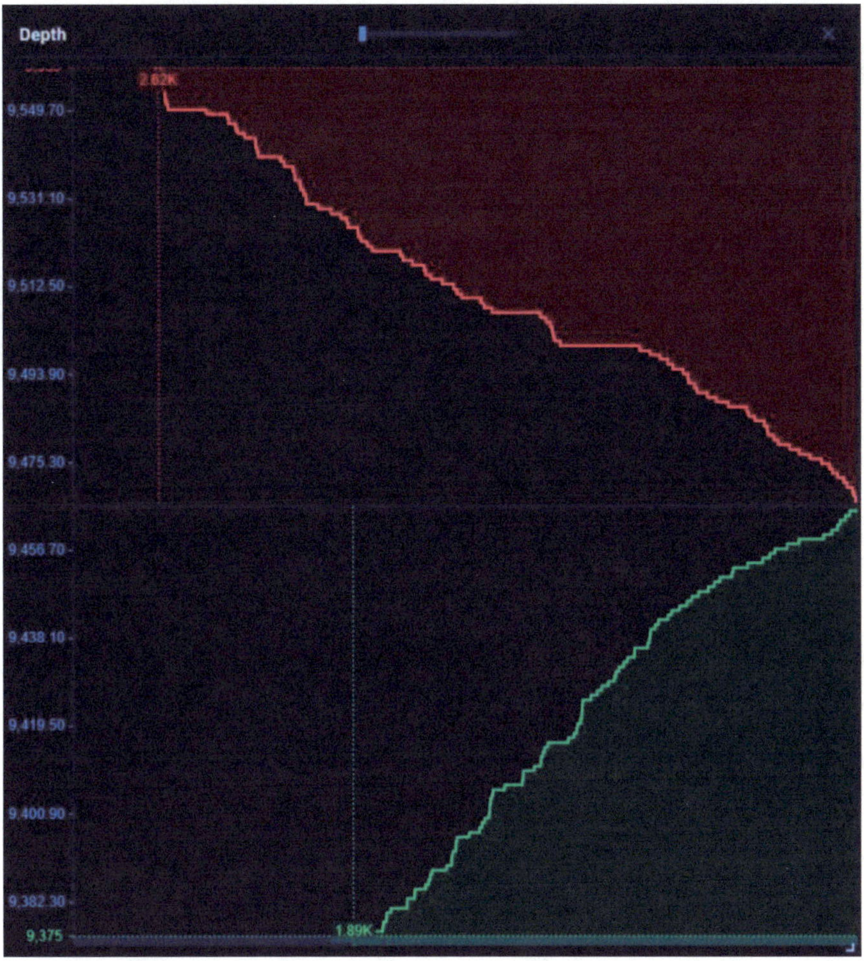

Order book depth of the BTC/USDT market pair.

In more general terms, the depth of the order book may also refer to the amount of liquidity that the order book can absorb. The "deeper" the market is, the more liquidity there is in the order book. In this sense, a market with more liquidity can absorb larger orders without a considerable effect on the price. However, if the market is illiquid, large orders may have a significant impact on the price.

What is a market order?

A market order is an order to buy or sell at the best currently available market price. It's basically the fastest way to get in or out of a market.

When you're setting a market order, you're basically saying: "I'd like to execute this order right now at the best price I can get."

Your market order will keep filling orders from the order book until the entire order is fully filled. This is why large traders (or whales) can have a significant impact on the price when they use market orders. A large market order can effectively siphon liquidity from the order book. How so? Let's go through it when discussing slippage.

What is slippage in trading?

There is something you need to be aware of when it comes to market orders – **slippage**. When we say that market orders fill at the best available price, that means that they keep filling orders from the order book until the entire order is executed.

However, what if there isn't enough liquidity around the desired price to fill a large market order? There could be a big difference between the price that you expect your order to fill and the price that it fills at. This difference is called slippage.

Let's say you'd like to open a long position worth 10 BTC in an altcoin. However, this altcoin has a relatively small market cap and is being traded on a low-liquidity market. If you use a market order, it will keep filling orders from the order book until the entire 10 BTC order is filled. On a liquid market, you would be able to fill your 10 BTC order without impacting the price significantly. But, in this case, the lack of liquidity means that there may not be enough sell orders in the order book for the current price range.

So, by the time the entire 10 BTC order is filled, you may find out that the average price paid was much higher than expected. In other words, the lack of sell orders caused your market order to move up the order book, matching orders that were significantly more expensive than the initial price.

Be aware of slippage when trading altcoins, as some trading pairs may not have enough liquidity to fill your market orders.

What is a limit order?

A limit order is an order to buy or sell an asset at a specific price or better. This price is called the **limit price**. Limit buy orders will execute at the limit price or lower, while limit sell orders will execute at the limit price or higher.

When you're setting a limit order, you're basically saying: "I'd like to execute this order at this specific price or better, but never worse."

Using a limit order allows you to have more control over your entry or exit for a given market. In fact, it guarantees that your order will never fill at a worse price than your desired price. However, that also comes with a downside. The market may never reach your price, leaving your order unfilled. In many cases, this can mean losing out on a potential trade opportunity.

Deciding when to use a limit order or market order can vary with each trader. Some traders may use only one or the other, while other traders will use both – depending on the circumstances. The important thing is to understand how they work so you can decide for yourself.

What is a stop-loss order?

Now that we know what market and limit orders are, let's talk about stop-loss orders. A stop-loss order is a type of limit or market order that's only activated when a certain price is reached. This price is called the **stop price**.

The purpose of a stop-loss order is mainly to limit losses. Every trade needs to have an invalidation point, which is a price level that you should define in advance. This is the level where you say that your initial idea was wrong, meaning that you should exit the market to prevent further losses. So, the invalidation point is where you would typically put your stop-loss order.

How does a stop-loss order work? As we've mentioned, the stop-loss can be both a limit or a market order. This is why these variants may also be referred to as stop-limit and stop-market orders. The key thing to understand is that the stop-loss only activates when a certain price is reached (the stop price). When the stop price is reached, it activates either a market or a limit order. You basically set the stop price as the trigger for your market or limit order.

However, there is one thing you should keep in mind. We know that limit orders only fill at the limit price or better, but never worse. If you're using a stop-limit order as your stop-loss and the market crashes violently, it may instantly move away from your limit price, leaving your order unfilled. In other words, the stop price would trigger your stop-limit order, but the limit order would remain unfilled due to the sharp price drop. This is why stop-market orders are considered safer than stop-limit orders. They ensure that even under extreme market conditions, you'll be guaranteed to exit the market once your invalidation point is reached.

What are makers and takers?

You become a **maker** when you place an order that doesn't immediately get filled but gets added to the order book. Since your order is adding liquidity to the order book, you're a "maker" of liquidity.

Limit orders will typically execute as maker orders, but not in all cases. For example, let's say you place a limit buy order with a limit price that's considerably higher than the current market price. Since you're saying your order can execute at the limit price or better, your order will execute against the market price (as it's lower than your limit price).

You become a **taker** when you place an order that gets immediately filled. Your order doesn't get added to the order book, but is immediately matched with an existing order in the order book. Since you're taking liquidity from the order book, you're a taker. Market orders will always be taker orders, as you're executing your order at the best currently available market price.

Some exchanges adopt a multi-tier fee model to incentivize traders to provide liquidity. After all, it's in their interest to attract high volume traders to their exchange – liquidity attracts more liquidity. In such systems, makers tend to pay lower fees than takers, since they're the ones adding liquidity to the exchange. In some cases, they may even offer fee rebates to makers.

What is the bid-ask spread?

The bid-ask spread is the difference between the highest buy order (bid) and the lowest sell order (ask) for a given market. It's essentially the gap between the highest price where a seller is willing to sell and the lowest price where a buyer is willing to buy.

The bid-ask spread is a way to measure a market's liquidity. The smaller the bid-ask spread is, the more liquid the market is. The bid-ask spread can also be considered as a measure of supply and demand for a given asset. In this sense, the supply is represented by the ask side while the demand by the bid side.

When you're placing a market buy order, it will fill at the lowest available ask price. Conversely, when you place a market sell order, it will fill at the highest available bid.

Chapter 7

Hopping Over the CandleStick

What is a candlestick chart?

A candlestick chart is a type of financial chart that graphically represents the price moves of an asset for a given timeframe. As the name suggests, it's made up of candlesticks, each representing the same amount of time. The candlesticks can represent virtually any period, from seconds to years.

Candlestick charts date back to about the 17th century. Their creation as a charting tool is often credited to a Japanese rice trader called Homma. His ideas were likely what provided the foundation for what is now used as the modern candlestick chart. Homma's findings were refined by many, most notably by Charles Dow, one of the fathers of modern technical analysis.

While candlestick charts could be used to analyze any other types of data, they are mostly employed to facilitate the analysis of financial markets. Used correctly, they're tools that can help traders gauge the probability of outcomes in the price movement. They can be useful as they enable traders and investors to form their own ideas based on their analysis of the market.

How do candlestick charts work?

The following price points are needed to create each candlestick:

- **Open** — The **first** recorded trading price of the asset within that particular timeframe.
- **High** — The **highest** recorded trading price of the asset within that particular timeframe.
- **Low** — The **lowest** recorded trading price of the asset within that particular timeframe.
- **Close** — The **last** recorded trading price of the asset within that particular timeframe.

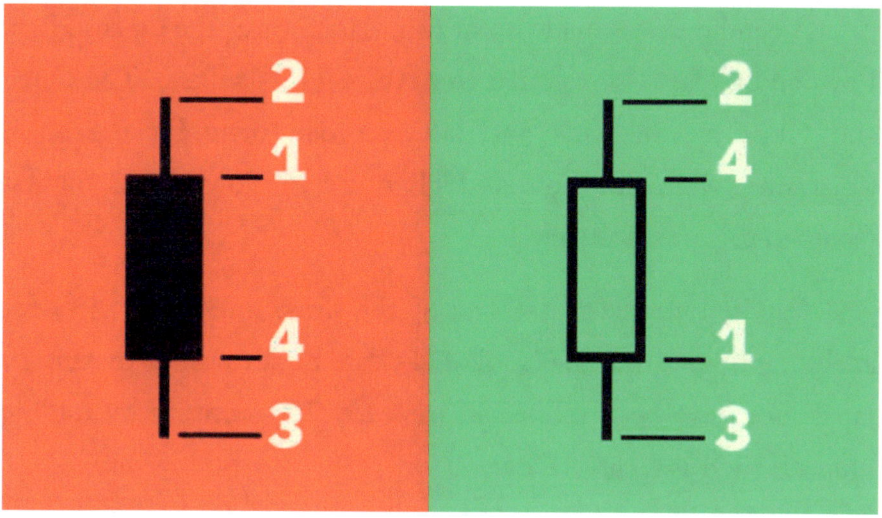

Collectively, this data set is often referred to as the OHLC values. The relationship between the open, high, low, and close determines how the candlestick looks.

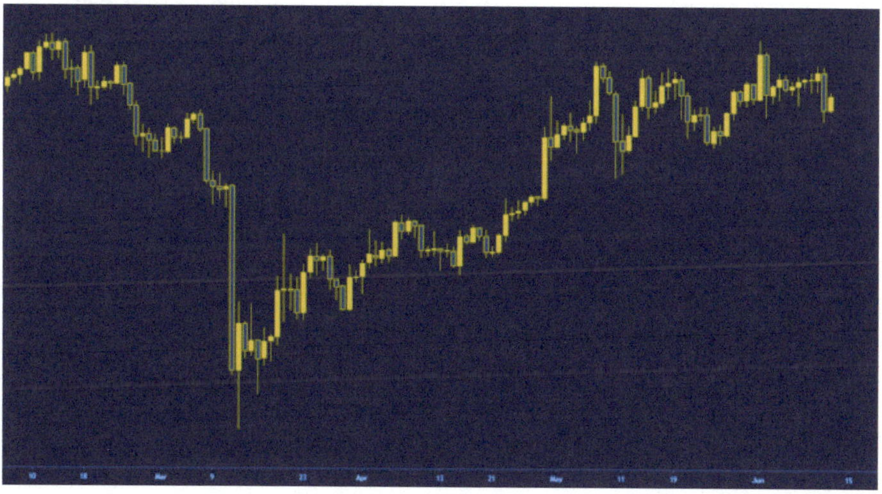

Daily chart of Bitcoin. Each candlestick represents one day of trading.

A candlestick is made up of four data points: the Open, High, Low, and Close (also referred to as the OHLC values). The Open and Close are the first and last recorded price for the given timeframe, while the Low and High are the lowest and highest recorded price, respectively.

Candlestick charts are one of the most important tools for analyzing financial data. Candlesticks date back to the 17th century Japan but have been refined in the early 20th century by trading pioneers such as Charles Dow.

Candlestick chart analysis is one of the most common ways to look at the Bitcoin market using technical analysis.

What is a candlestick chart pattern?

Technical analysis is largely based on the assumption that previous price movements may indicate future price action. So, how can candlesticks be useful in this context? The idea is to identify candlestick chart patterns and create trade ideas based on them.

Candlestick charts help traders analyze market structure and determine whether we're in a bullish or bearish market environment. They may also be used to identify areas of interest on a chart, like support or resistance levels or potential points of reversal. These are the places on the chart that usually have increased trading activity.

Candlestick patterns are also a great way to manage risk, as they can present trade setups that are defined and exact. How so? Well, candlestick patterns can define clear price targets and invalidation points. This allows traders to come up with very precise and controlled trade setups. As such, candlestick patterns are widely used by Forex and cryptocurrency traders alike.

The distance between the open and close is referred to as the body, while the distance between the body and the high/low is referred to as the wick or shadow. The distance between the high and low of the candle is called the range of the candlestick.

How to read candlestick charts

Many traders consider candlestick charts easier to read than the more conventional bar and line charts, even though they provide similar information. Candlestick charts can be read at a glance, offering a simple representation of price action.

In practice, a candlestick shows the battle between bulls and bears for a certain period. Generally, the longer the body is, the more intense the buying or selling pressure was during the measured timeframe. If the wicks on the candle are short, it means that the high (or the low) of the measured timeframe was near the closing price.

The color and settings may vary with different charting tools, but generally, if the body is green, it means that the asset closed higher than it opened. Red means that the price moved down during the measured timeframe, so the close was lower than the open.

Some chartists prefer to use black-and-white representations. So instead of using green and red, the charts represent up movements with hollow candles and down moves with black candles. Candlestick charts are one of the most commonly used technical tools to analyze price patterns. They have been used by traders and investors for centuries to find patterns that may indicate where the price is headed. This article will cover some of the most well-known candlestick patterns with illustrated examples.

How to use Candlestick Patterns

There are countless candlestick patterns that traders can use to identify areas of interest on a chart. These can be used for day trading, swing trading, and even longer-term position trading. While some candlestick patterns may provide insights into the balance between buyers and sellers, others may indicate a reversal, continuation, or indecision.

It's important to note that candlestick patterns aren't necessarily a buy or sell signal by themselves. They are instead a way to look at market structure and a potential indication of an upcoming opportunity. As such, it is always useful to look at patterns in context. This can be the context of the technical pattern on the chart, but also the broader market environment and other factors.

In short, like any other market analysis tool, candlestick patterns are most useful when used in combination with other techniques. These may include the Wyckoff Method, the Elliott Wave Theory and the Dow Theory. It can also include technical analysis (TA) indicators, such as Trend Lines, Moving Averages, the Relative Strength Index (RSI), Stochastic RSI, Bollinger Bands, Ichimoku Clouds, Parabolic SAR, or the MACD.

BULLISH REVERSAL PATTERNS

Hammer

A candlestick with a long lower wick at the bottom of a downtrend, where the lower wick is at least twice the size of the body.

A hammer shows that even though the selling pressure was high, the bulls drove the price back up close to the open. A hammer can be either red or green, but green hammers may indicate a stronger bull reaction.

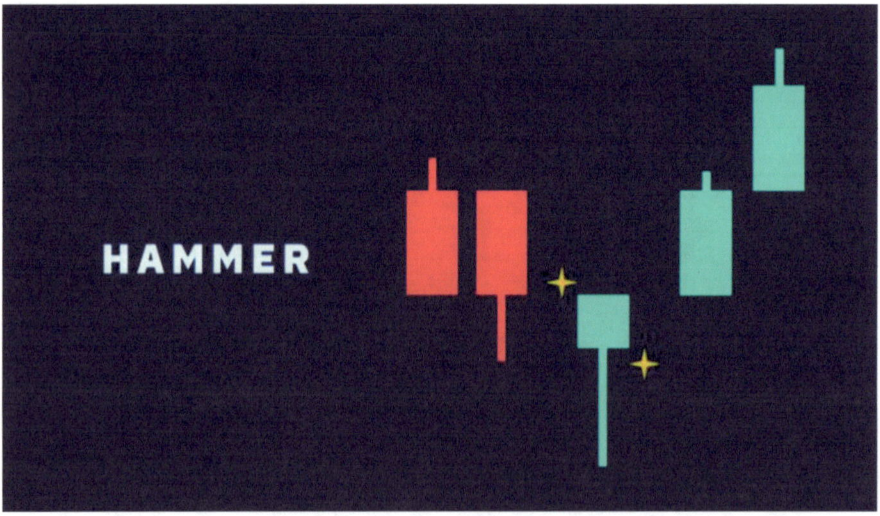

Inverted hammer

Also called the inverse hammer, it's just like a hammer, but with a long wick above the body rather than below. Similar to a hammer, the upper wick should be at least twice the size of the body.

An inverted hammer occurs at the bottom of a downtrend and may indicate a potential reversal upward. The upper wick shows that price stopped its continued downward movement, even though the sellers eventually managed to drive it down near the open. As such, the inverted hammer may suggest that buyers soon might gain control of the market.

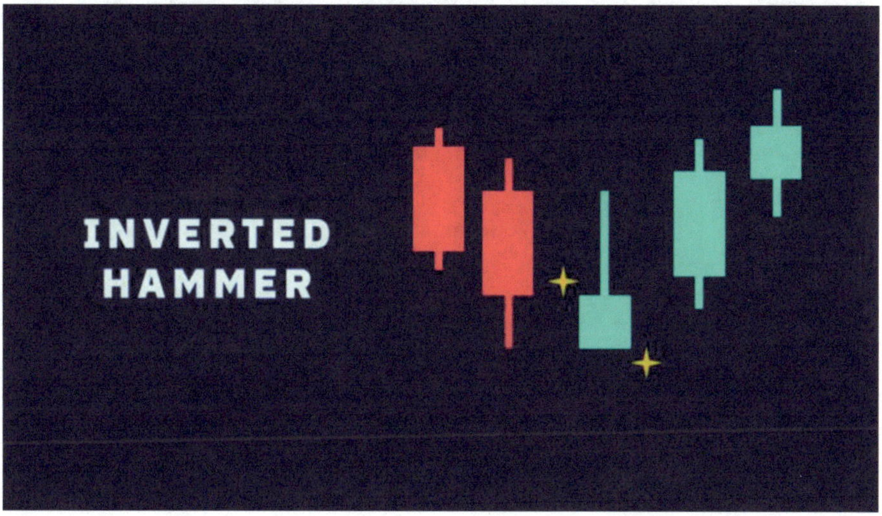

Three White Soldiers

The three white soldiers pattern consists of three consecutive green candlesticks that all open within the previous candle's body, and close at a level exceeding the previous candle's high.

Ideally, these candlesticks shouldn't have long lower wicks, indicating that continuous buying pressure is driving the price up. The size of the candles and the length of the wicks can be used to judge the chances of continuation or a possible retracement.

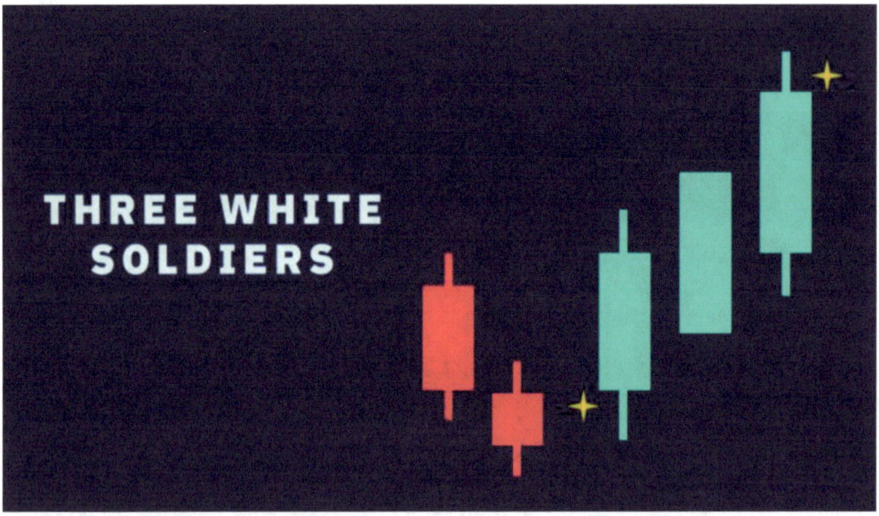

Bullish Harami

A bullish harami is a long red candle followed by a smaller green candle that's entirely contained within the body of the previous candle.

The bullish harami can unfold over two or more days, and it's a pattern indicating that selling momentum is slowing down and might be coming to an end.

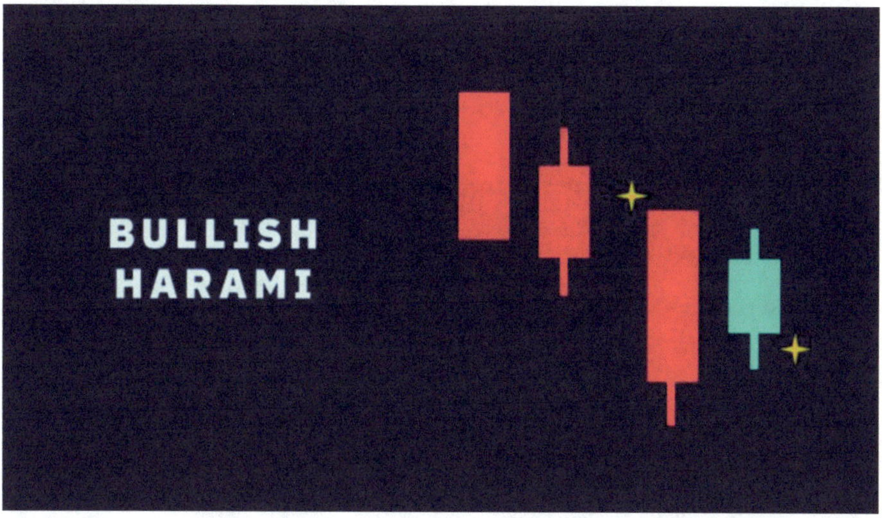

BEARISH REVERSAL PATTERNS

Hanging Man

The hanging man is the bearish equivalent of a hammer. It typically forms at the end of an uptrend with a small body and a long lower wick.

The lower wick indicates that there was a large sell-off, but bulls managed to take back control and drive the price up. Keeping that in mind, after a prolonged uptrend, the sell-off may act as a warning that the bulls might soon be losing control of the market.

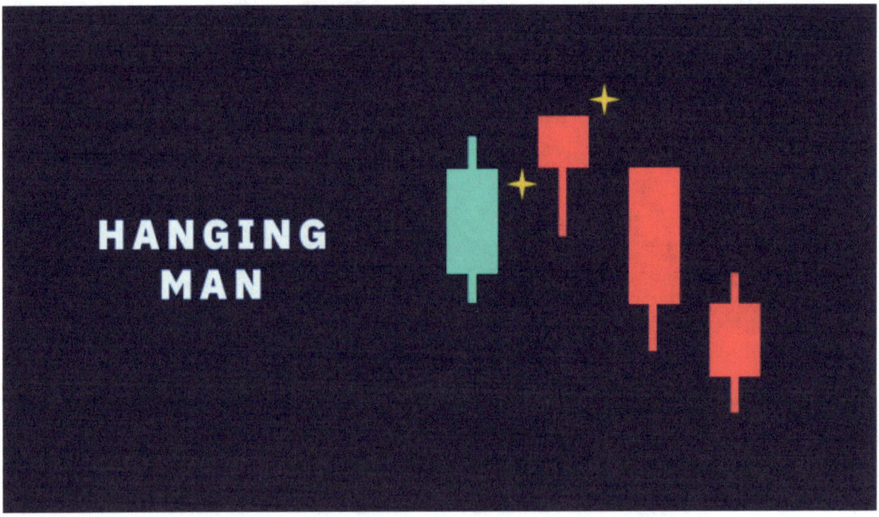

Shooting Star

The shooting star is made of a candlestick with a long upper wick, little or no lower wick, and a small body, ideally near the low. The shooting star is a similar shape as the inverted hammer but is formed at the end of an uptrend.

It indicates that the market reached a high, but then sellers took control and drove the price back down. Some traders prefer to wait for the next few candlesticks to unfold for confirmation of the pattern.

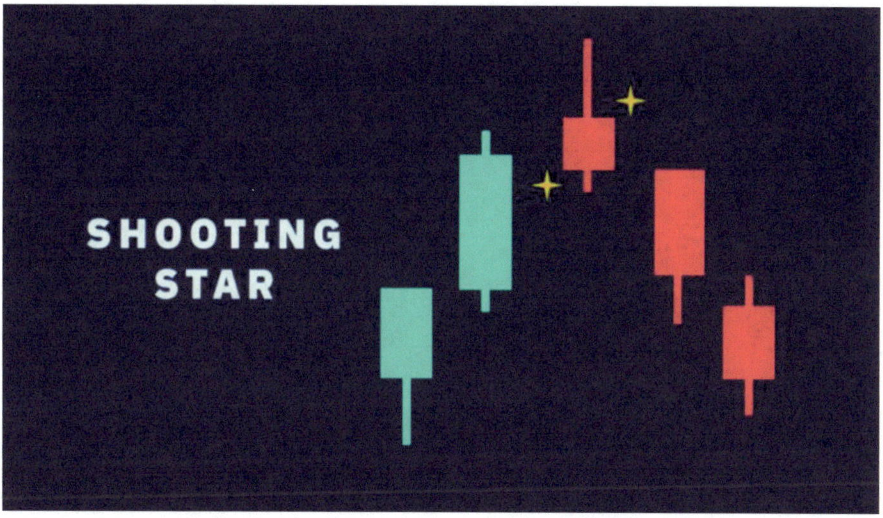

Three Black Crows

The three black crows are made of three consecutive red candlesticks that open within the previous candle's body, and close at a level below the previous candle's low.

The bearish equivalent of three white soldiers. Ideally, these candlesticks shouldn't have long higher wicks, indicating continuous selling pressure driving the price down. The size of the candles and the length of the wicks can be used to judge the chances of continuation.

Bearish Harami

The bearish harami is a long green candle followed by a small red candle with a body that's entirely contained within the body of the previous candle.

The bearish harami can unfold over two or more days, appears at the end of an uptrend, and may indicate that buying pressure is decreasing.

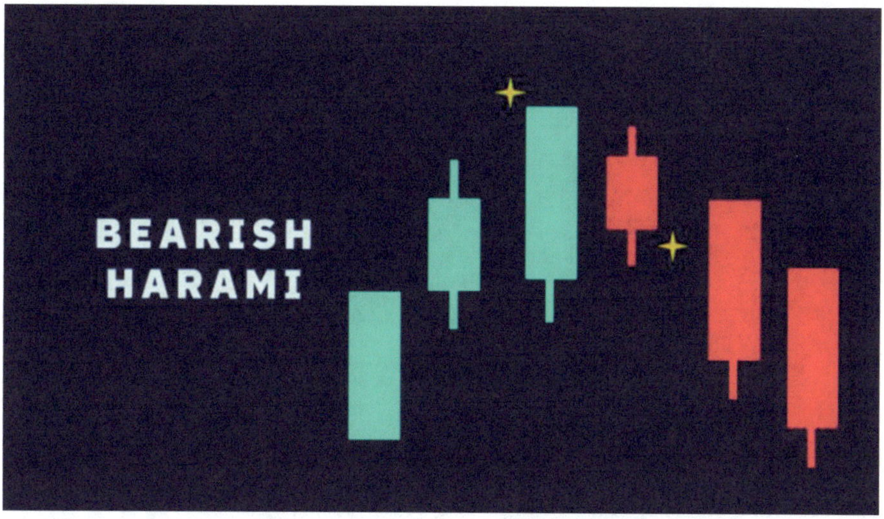

Dark Cloud Cover

The dark cloud cover pattern consists of a red candle that opens above the close of the previous green candle but then closes below the midpoint of that candle.

It can often be accompanied by high volume, indicating that momentum might be shifting from the upside to the downside. Traders might wait for a third red candle for confirmation of the pattern.

Rising Three Methods

This pattern occurs in an uptrend, where three consecutive red candles with small bodies are followed by the continuation of the uptrend. Ideally, the red candles shouldn't breach the range of the preceding candlestick.

The continuation is confirmed with a green candle with a large body, indicating that bulls are back in control of the trend's direction.

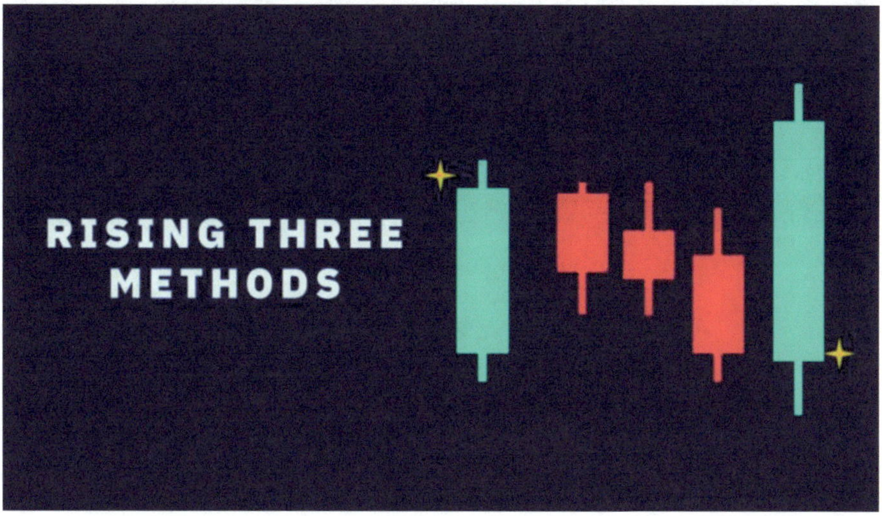

Falling Three Methods

The inverse of rising three methods, indicating the continuation of a downtrend instead.

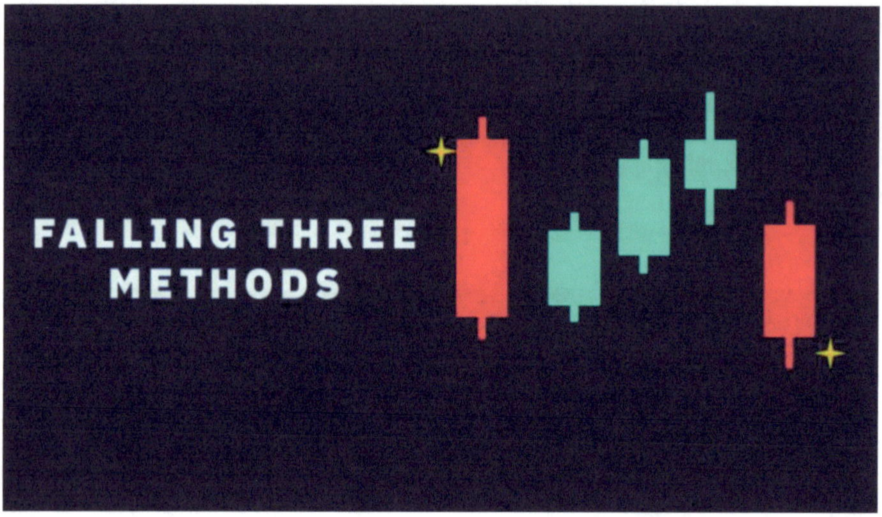

Doji

A Doji forms when the open and the close are the same (or very close to each other). The price can move above and below the open but eventually closes at or near the open. As such, a Doji may indicate an indecision point between buying and selling forces. Still, the interpretation of a Doji is highly dependent on context.

Depending on where the line of the open/close falls, a Doji can be described as:

Gravestone Doji – Bearish reversal candle with a long upper wick and the open/close near the low.

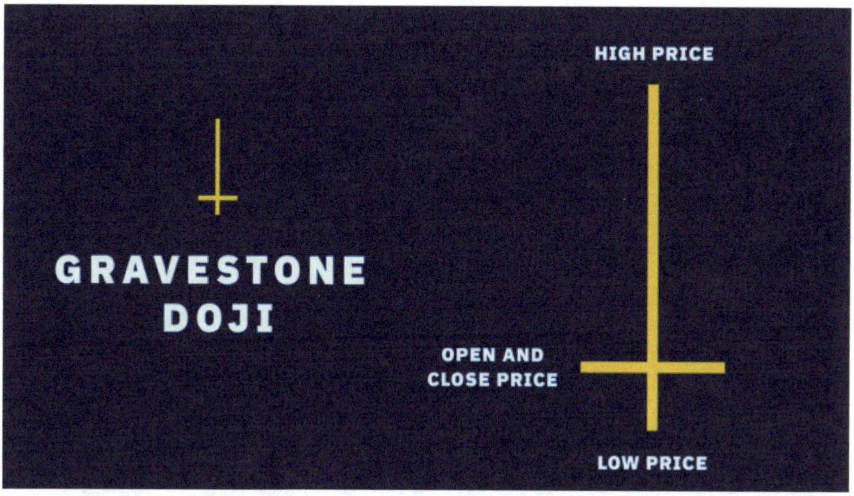

Long-legged Doji – Indecisive candle with both a lower and upper wick, and the open/close near the midpoint.

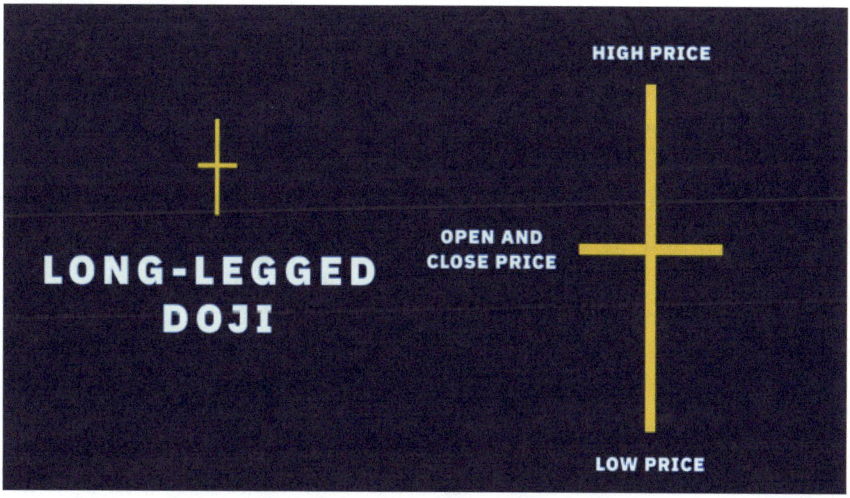

Dragonfly Doji – Either bullish or bearish candle (depending on context) with a long lower wick and the open/close near the high.

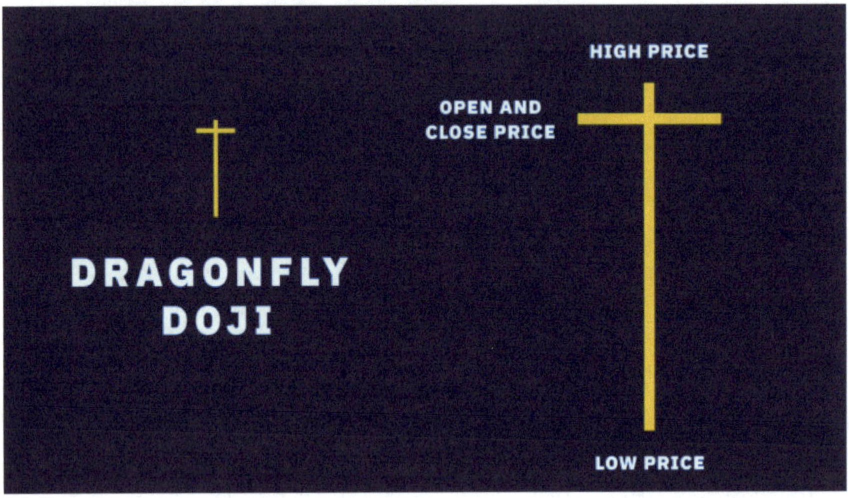

According to the original definition of the Doji, the open and close should be exactly the same. But what if the open and close aren't the same but are instead very close to each other? That's called a spinning top. However, since cryptocurrency markets can be very volatile, an exact Doji is rare. As such, the spinning top is often used interchangeably with the Doji.

Candlestick patterns based on price gaps

There are many candlestick patterns that use price gaps. A price gap is formed when a financial asset opens above or below its previous closing price, which creates a gap between the two candlesticks. Since cryptocurrency markets trade round the clock, patterns based on these types of price gaps are not present. Even so, price gaps can still occur in illiquid markets. However, since they happen mainly because of low liquidity and high bid-ask spreads, they might not be useful as actionable patterns.

Candlestick patterns are essential for any trader to at least be familiar with, even if they don't directly incorporate them into their trading strategy.

While they can be undoubtedly useful to analyze the markets, it's important to remember that they aren't based on any scientific principles or laws. They instead convey and visualize the buying and selling forces that ultimately drive the markets.

What candlestick charts don't tell you

While candlesticks are useful in giving you a general idea of price action, they may not provide all you need for a comprehensive analysis. For instance, candlesticks don't show in detail what happened in the interval between the open and close, only the distance between the two points (along with the highest and lowest prices).

For example, while the wicks of a candlestick do tell us the high and low of the period, they can't tell us which one happened first. Still, in most charting tools, the timeframe can be changed, allowing traders to zoom into lower timeframes for more details.

Candlestick charts can also contain a lot of market noise, especially when charting lower timeframes. The candles can change very quickly, which can make them challenging to interpret.

Heikin-Ashi Candlesticks

So far, we have discussed what is sometimes referred to as the Japanese candlestick chart. But there are other ways to calculate candlesticks. The Heikin-Ashi Technique is one of them.

Heikin-Ashi stands for "average bar" in Japanese. Such candlestick charts rely on a modified formula that uses average price data. The main goal is to smooth out price action and filter out market noise. As such, Heikin-Ashi candles can make it easier to spot market trends, price patterns, and possible reversals.

Traders often use Heikin-Ashi candles in combination with Japanese candlesticks to avoid false signals and increase the chances of spotting market trends. Green Heikin-Ashi candles with no lower wicks generally indicate a strong uptrend, while red candles with no upper wicks may point to a strong downtrend.

While Heikin-Ashi candlesticks can be a powerful tool, like any other technical analysis technique, they do have their limitations. Since these candles use averaged price data, patterns may take longer to develop. Also, they don't show price gaps and may obscure other price data.

Candlestick charts are one of the most fundamental tools for any trader or investor. They not only provide a visual representation of the price action for a given asset, but also offer the flexibility to analyze data in different timeframes. practice may eventually provide traders with an edge over the market. Still, most traders and investors agree that it's also important to consider other methods, such as fundamental analysis.

An extensive study of candlestick charts and patterns, combined with an analytical mindset and enough

What is a trend line?

Trend lines are a widely used tool by both traders and technical analysts. They are lines that connect certain data points on a chart. Typically, this data is the price, but not in all cases. Some traders may also draw trend lines on technical indicators and oscillators.

The main idea behind drawing trend lines is to visualize certain aspects of the price action. This way, traders can identify the overall trend and market structure.

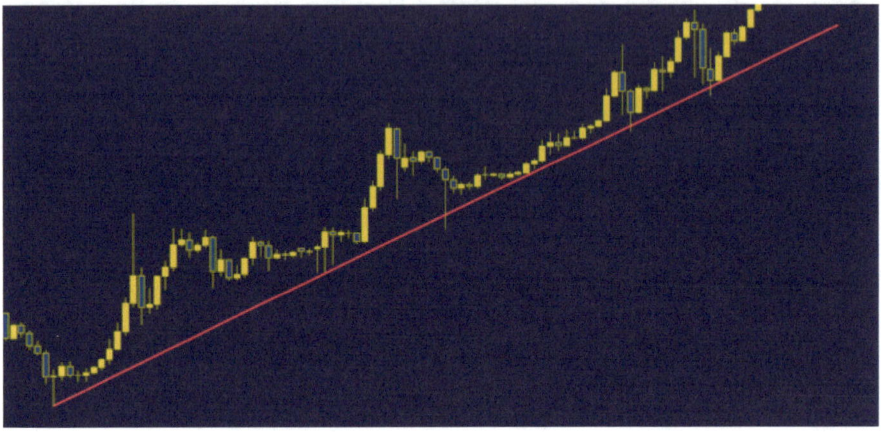

The price of Bitcoin touching a trend line multiple times, indicating an uptrend.

Some traders may only use trend lines to get a better understanding of the market structure. Others may use them to create actionable trade ideas based on how the trend lines interact with the price.

Trend lines can be applied to a chart showing virtually any time frame. However, as with any other market analysis tool, trend lines on higher time frames tend to be more reliable than trend lines on lower time frames.

Another aspect to consider here is the strength of a trend line. The conventional definition of a trend line defines that it has to touch the price at least two or three times to become valid. Typically, the more times the price has touched (*tested*) a trend line, the more reliable it may be considered.

What are support and resistance?

Support and resistance are some of the most basic concepts related to trading and technical analysis.

Support means a level where the price finds a "floor." In other words, a support level is an area of significant demand, where buyers step in and push the price up.

Resistance means a level where the price finds a "ceiling." A resistance level is an area of significant supply, where sellers step in and push the price down.

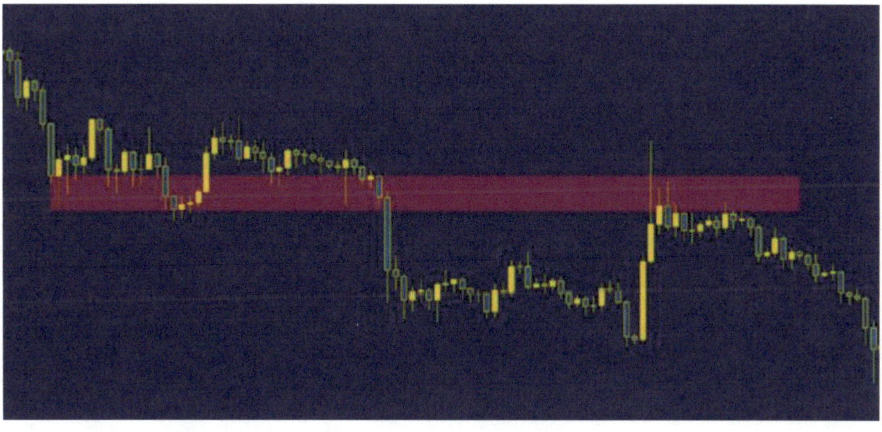

Support level (red) is tested and broken, turning into resistance.

Now you know that support and resistance are levels of increased demand and supply, respectively. However, many other factors can be at play when thinking about support and resistance.

Technical indicators, such as trend lines, moving averages, Bollinger Bands, Ichimoku Clouds, and Fibonacci Retracement can also suggest potential support and resistance levels. In fact, even aspects of human psychology are used. This is why traders and investors may incorporate support and resistance very differently in their individual trading strategy.

Chapter 8

Charting New Territory

How do I start trading cryptocurrency?

If you've decided you'd like to start trading, here are a few things to consider.

Firstly, you'll, of course, need capital to trade with. If you don't have savings and start trading with money you can't lose, it can have a seriously detrimental impact on your life. Trading isn't an easy feat – an overwhelming majority of beginner traders lose money. You'll need to expect that the money you put aside for trading can vanish quickly, and you may never recover your losses. This is why it's recommended to start with smaller amounts to test out the waters.

Something else you'll also need to think about is your overall trading strategy. There are a lot of possible avenues to take when it comes to making money in the financial markets. Depending on the time and effort you can put into this undertaking, you can choose between many different strategies to achieve your financial goals.

Lastly, here's an additional point. Many traders are at their best when trading isn't their main source of income. This way, the emotional burden is easier to bear than if their day-to-day survival depended on it. Eliminating emotion is a core trait of successful traders, and it's significantly harder to do when one's livelihood is at stake. So, especially when you're starting out, you could think of trading and investing as a side venture. And remember to start with small amounts for the sake of learning and practicing. It may also be beneficial to look into ways of making passive income with cryptocurrency.numbers, and if you're not doing something well, that will be reflected in your performance. By meticulously keeping a trading journal, you can also monitor what strategies perform best.

How should I calculate my position size in trading?

One of the most important aspects of trading is risk management. In fact, some traders argue it is *the* most important thing. This is why it's critical to calculate the size of your positions with a standardized formula. Here's how the calculation goes.

First, you need to determine how much of your account you are willing to risk on individual trades. Let's say this is 1%. Does it mean you enter positions with 1% of your account? No, it means that if your stop-loss is hit, you won't lose more than 1% of your account.

That may seem too little, but this is to make sure that a few inevitable bad trades won't blow up your account. So, once you've got this defined, you need to determine where your stop-loss is. You do this for each individual trade, based on the specifics of the trade idea. Let's say you've determined that you're going to place your stop-loss 5% from your initial entry. This means that when your stop-loss is hit, and you exit 5% from your entry, you should lose exactly 1% of your account.

So, let's say our account size is 1000 USDT. We're risking 1% with each trade. Our stop-loss is 5% from our entry. What position size should we use?

1000*0.01/0.05=200

If we want to only lose 10 USDT, which is 1% of our account, we should enter a 200 USDT position.

What platforms can I trade on?

Chart analysis is a core part of any technical analyst's trading toolkit. But where is the best way to do it? Binance has integrated TradingView charts, so you can do your analysis directly on the platform – both on the web interface and in the mobile app. You can also create a TradingView account and check all Binance markets through their platform. Kucoin, Kraken, Gate.io, and Coinbase Pro are all good trading tools.

There are numerous other online charting software providers in the market, each providing different benefits. Typically, though, you'll have to pay a monthly subscription fee. Some other ones focused on crypto trading are Coinigy, TradingLite, Exocharts, and Tensorcharts.

What is a pump and dump (P&D)?

A pump and dump are a scheme that involves boosting the price of an asset through false information. When the price has gone up a significant amount ("pumped"), the perpetrators sell ("dump") their cheaply bought bags at a much higher price.

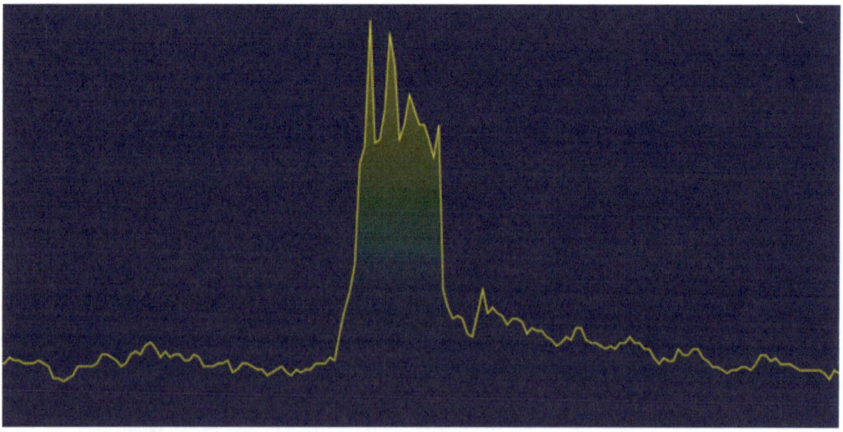

Typical price pattern of a pump and dump scheme.

Pump and dump schemes are rampant in the cryptocurrency markets, especially in bull markets. During these times, many inexperienced investors enter the market, and they are easier to take advantage of. This type of fraud is most common with small market cap cryptocurrencies, as their prices are generally easier to inflate due to the low liquidity of these markets.

Pump and dump schemes are often orchestrated by private "pump and dump groups" that promise easy returns for joiners (usually in exchange for a fee). However, what usually happens is that those joiners are taken advantage of by an even smaller group who have already built their positions.

In the legacy markets, people found guilty of facilitating pump and dump schemes are subject to hefty fines.

Crypto Airdrops

Airdrops are a novel way of distributing cryptocurrencies to a wide audience. An airdrop can be a great way to make sure that a cryptocurrency isn't centralized in the hands of only a few holders. A diverse set of holders is paramount for a healthy, decentralized network.

However, there's no such thing as a free lunch. Well, sometimes, there may be, if you get very lucky! Typically, though, what happens is that the promoters of the airdrop will outright try to take advantage of you, or will want something in return.

What will they ask for? One of the most common "assets" asked in return for an airdrop is your personal information. Is your personal data worth $10-50 worth of a highly speculative cryptocurrency? That's your choice to make, but there may be better ways to earn a bit of side income, without putting your privacy and personal data at risk. This is why you need to be extra careful when thinking about signing up for cryptocurrency airdrops.

What Is Technical Analysis?

Technical analysis also referred to as charting helps to predict how the markets behave based on previous volume data and price action. Traditionally, this has a lot to do with the stock markets but can also be applied to the crypto markets. This is how you get familiar with identifying tends and trading opportunities.

Also, TA is the study of an asset's current and previous prices. To examine prices and spot good opportunities, traders use charting tools known as indicators. It's good to use these indicators to identify existing trends and also provide more information about trends that could arise in the future. It's always a good choice to learn how to use multiple indicators to keep your risk at a minimum.

What is Fundamental Analysis (FA)?

Fundamental Analysis takes a look at an asset's valuation by studying the financial factors to see if the value is far. Certain situations like industry conditions or the companies connects to the assets are included. When the analysis is done, an analysis makes a decision on if an asset is overvalued or undervalued. When it comes to the crypto markets there is a new field of data science with public blockchain data called on-chain metrics. These metrics include the network hash rate, the top holders, the number of addresses, and analysis of transactions. Due to crypto's new entry into the world, at this time, fundamental analysis is less suitable. As the market matures, valuations for may be more accurate

What is a Momentum Indicator?

Momentum indicators help to show traders when identify good entry points for day trader's and scalpers on a short-term timeframe. If volatility is low in the market, then the prices will shrink into a small range. As the tensions fire up, the price makes a big jump, breaking out of the range. When the move is done, traders exit their positions and look for the next big momentum push.

What is the Trading Volume?

The trading volume may be considered the quintessential indicator. It shows the number of individual units traded for an asset in a given time. It basically shows how much of that asset changed hands during the measured time.

One of the best indicators considered is trading volume. Volume identifies the number of individual units traded at any given time. This gives insight into how much is being exchanged over a time period.

By using volume in trading, traders can measure the strength of the underlying trend. If high volatility is accompanied by high trading volume, that may be considered a validation of the move. This makes sense because high trading activity should equal a significant volume since many traders and investors are active at that particular price level. However, if volatility isn't accompanied by high volume, the underlying trend may be considered weak.

Price levels with historically high volume may also give a good potential entry or exit point for traders. Since history tends to repeat itself, these levels may be where increased trading activity is more likely to happen. Ideally, support and resistance levels should also be accompanied by an uptick in volume, confirming the strength of the level.

What is the Relative Strength Index (RSI)?

The RSI indicators will show if an asset is overbought or oversold. This oscillator varies between 0 and 100 and based on momentum can show you at which price changes happen.

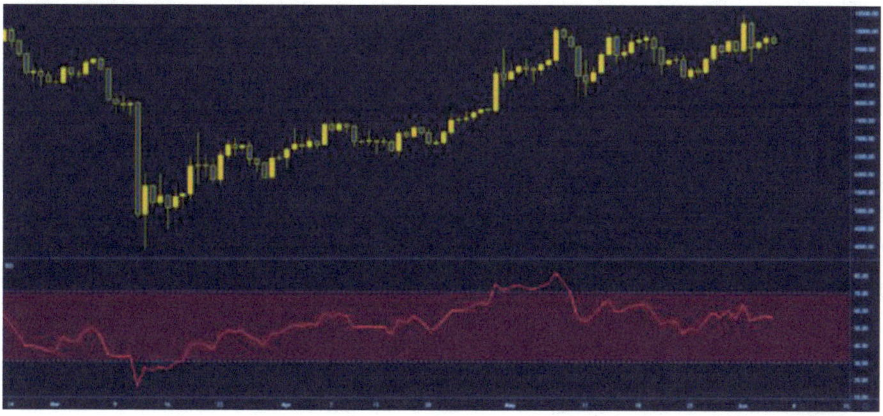

The RSI indicator applied to a Bitcoin chart.

If momentum is going up while the price is going up then we have a strong uptrend. On the other hand, if the momentum is losing gas in an uptrend, then it is potentially weak and a reversal may be on the way. When the RSI value is under 30, the asset may be considered oversold. In contrast, it may be considered overbought when it's above 70.

What is a Moving Average (MA)?

Moving averages smooth out price action and make it easier to spot market trends. As they're based on previous price data, they lack predictive qualities. As such, moving averages are considered lagging indicators.

Moving averages have various types – the two most common one is the **simple moving average** (SMA or MA) and the **exponential moving average** (EMA). What's the difference between them?

The simple moving average is calculated by taking price data from the previous *n* periods and producing an average. For example, the 10-day SMA takes the average price of the last 10 days and plots the results on a graph.

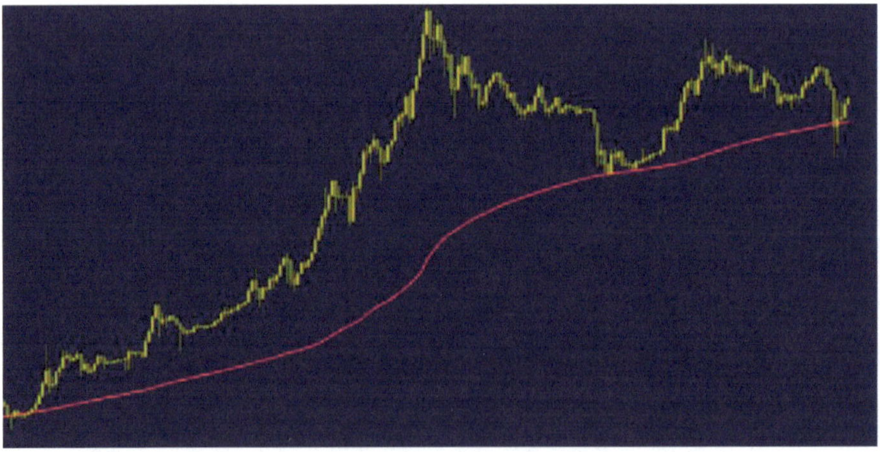

200-week moving average based on the price of Bitcoin.

The exponential moving average is a bit trickier. It uses a different formula that puts a bigger emphasis on more recent price data. As a result, the EMA reacts more quickly to recent events in price action, while the SMA may take more time to catch up.

As we've mentioned, moving averages are lagging indicators. The longer the period they plot, the greater the lag. As such, a 200-day moving average will react slower to unfolding price action than a 100-day moving average.

What is the Moving Average Convergence Divergence (MACD)?

The MACD is an oscillator that uses two moving averages to show the momentum of a market. As it tracks price action that has already occurred, it's a lagging indicator.

The MACD is made up of two lines – the MACD line and the signal line. How do you calculate them? Well, you get the MACD line by subtracting the 26 EMA from the 12 EMA. Simple enough. Then, you plot this over the MACD line's 9 EMA – the signal line. In addition, many charting tools will also show a histogram that illustrates the distance between the MACD line and the signal line.

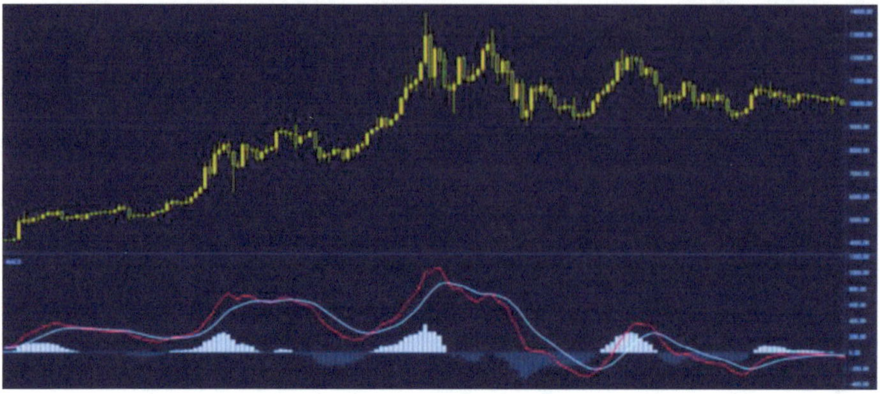

The MACD indicator applied to a Bitcoin chart.

Traders may use the MACD by observing the relationship between the MACD line and the signal line. A crossover between the two lines is usually a notable event when it comes to the MACD. If the MACD line crosses above the signal line, that may be interpreted as a bullish signal. In contrast, if the MACD line crosses below the signal, that may be interpreted as a bearish signal.

The MACD is one of the most popular technical indicators out there to measure market momentum. If you'd like to read more about it, check out *MACD Indicator Explained*.

What is the Fibonacci Retracement tool?

The Fibonacci Retracement (or Fib Retracement) tool is a popular indicator based on a string of numbers called the Fibonacci sequence. These numbers were identified in the 13th century, by an Italian mathematician called Leonardo Fibonacci.

The Fibonacci numbers are now part of many technical analysis indicators, and the Fib Retracement is among the most popular ones. It uses ratios derived from the Fibonacci numbers as percentages. These percentages are then plotted over a chart, and traders can use them to identify potential support and resistance levels.

These Fibonacci ratios are:

- 0%
- 23.6%
- 38.2%
- 61.8%
- 78.6%
- 100%

While 50% is technically not a Fibonacci ratio, many traders also consider it when using the tool. In addition, Fibonacci ratios outside of the 0-100% range may also be used. Some of the most common ones are 161.8%, 261.8%, and 423.6%.

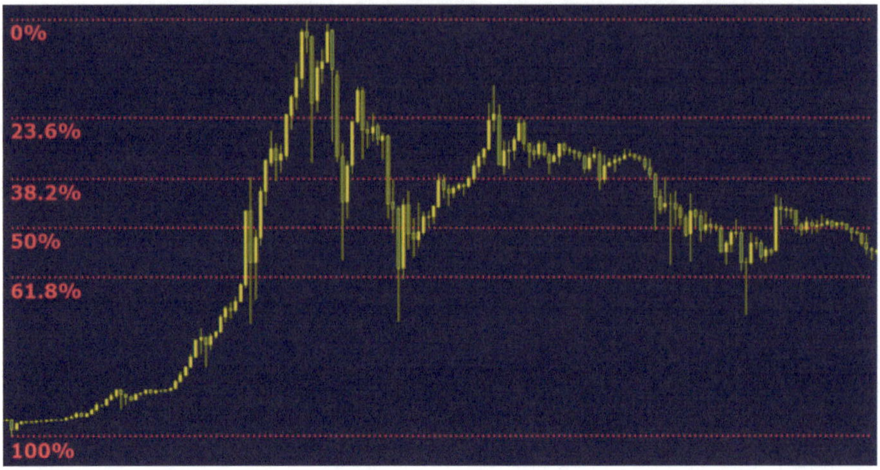

Fibonacci levels on a Bitcoin chart.

So, how can traders use the Fibonacci Retracement levels? The main idea behind plotting percentage ratios on a chart is to find areas of interest. Typically, traders will pick two significant price points on a chart, and pin the 0 and 100 values of the Fib Retracement tool to those points. The range outlined between these points may highlight potential entry and exit points, and help determine stop-loss placement.

The Fibonacci Retracement tool is a versatile indicator that can be used in a wide range of trading strategies. If you'd like to read more, check *A Guide to Mastering Fibonacci Retracement*.

What is the Stochastic RSI (StochRSI)?

The Stochastic RSI, or StochRSI, is a derivative of the RSI. Similarly, to the RSI, it's main goal is to determine whether an asset is overbought or oversold. In contrast to the RSI, however, the StochRSI isn't generated from price data, but RSI values. On most charting tools, the values of the StochRSI will range between 0 and 1 (or 0 and 100).

The StochRSI tends to be the most useful when it's near the upper or lower extremes of its range. However, due to its greater speed and higher sensitivity, it may produce a lot of false signals that can be challenging to interpret.

The traditional interpretation of the StochRSI is somewhat similar to that of the RSI. When it's over 0.8, the asset may be considered overbought. When it's below 0.2, the asset may be considered oversold. However, it's worth mentioning that these shouldn't be viewed as direct signals to enter or exit trades. While this information is certainly telling a story, there may be other sides to the story as well. This is why most technical analysis tools are best used in combination with other market analysis techniques.

What are Bollinger Bands (BB)?

Named after John Bollinger, the Bollinger Bands measure market volatility, and are often used to spot overbought and oversold conditions. This indicator is made up of three lines, or "bands" – an SMA (the middle band), and an upper and lower band. These bands are then placed on a chart, along with the price action. The idea is that as volatility increases or decreases, the distance between these bands will change, expanding and contracting.

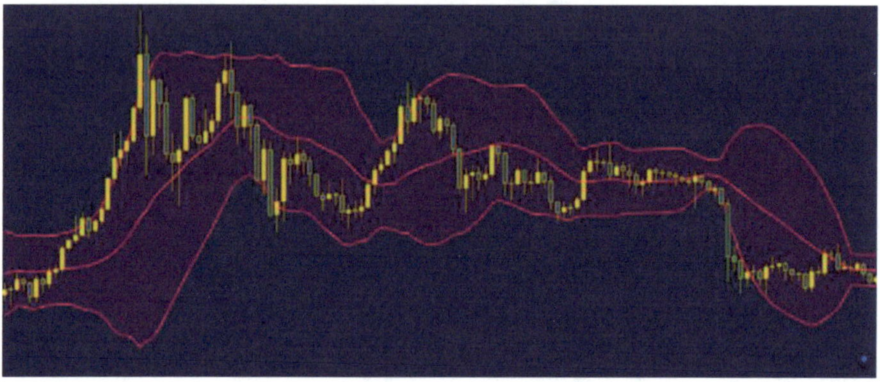

Bollinger Bands on a Bitcoin chart.

Let's go through the general interpretation of Bollinger Bands. The closer the price is to the upper band, the closer the asset may be to overbought conditions. Similarly, the closer it is to the lower band, the closer the asset may be to oversold conditions.

One thing to note is that the price will generally be contained within the range of the bands, but it may break above or below them at times. Does this mean that it's an immediate signal to buy or sell? No. It just tells us that the market is moving away from the middle band SMA, reaching extreme conditions.

Traders may also use Bollinger Bands to try and predict a market squeeze, also known as the Bollinger Bands Squeeze. This refers to a period of low volatility when the bands come really close to each other and "squeeze" the price into a small range. As the "pressure" builds up in that small range, the market eventually pops out of it, leading to a period of increased volatility. Since the market can move up or down, the squeeze strategy is considered neutral (neither bearish or bullish). So, it might be worth combining it with other trading tools, such as support and resistance.

What is the Volume-Weighted Average Price (VWAP)?

As we've discussed earlier, many traders consider the trading volume to be the most important indicator out there. So, are there any indicators based on volume?

The volume-weighted average price, or VWAP, combines the power of volume with price action. In more practical terms, it's the average price of an asset for a given period weighted by volume. This makes it more useful than simply calculating the average price, as it also takes into account which price levels had the most trading volume.

How do traders use the VWAP? Well, the VWAP is typically used as a benchmark for the current outlook on the market. In this sense, when the market is above the VWAP line, it may be considered bullish. At the same time, if the market is below the VWAP line, it may be considered bearish. Have you noticed how this is similar to the interpretation of moving averages? The VWAP may indeed be compared to moving averages, at least in the way it's used. As we've seen, the main difference is that the VWAP considers the trading volume as well.

In addition, the VWAP can also be used to identify areas of higher liquidity. Many traders will use the price breaking above or below the VWAP line as a trade signal. However, they will typically also incorporate other metrics into their strategy to reduce risks.

The most fundamental indicator is volume. Volume is used as a tool for confirming a trend or identifying potential points of reversal. The VWAP combines volume & price to create a practical indicator. Traders rely on the VWAP as a tool for trend confirmation to identify entry and exit points.

Breathe in The VWAP

VWAP stands for **volume-weighted average price**. It's the average price of the asset for a given period weighted by volume.

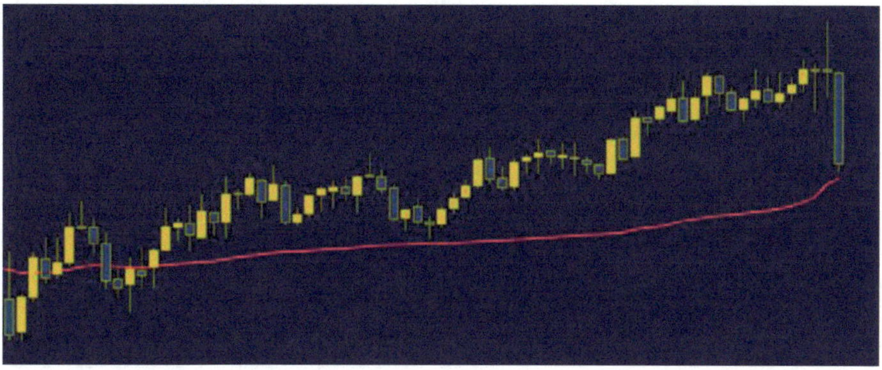

The VWAP incorporates volume into the average price calculation. The combines these two important metrics into one indicator. The VWAP can give an indication of a dominant market trend and important areas of liquidity.

Exhale The Noise

The VWAP may be used as a benchmark for a more passive, longer-term investment style in the current market outlook. For example, a simple strategy may be only to buy assets that are below their VWAP line, indicating that they're potentially undervalued.

With that said, some traders may use the price crossing the VWAP line as a signal to enter a trade. If the price breaches and goes above the VWAP, they may get into a long position. Conversely, if the price breaches and goes below the VWAP, they may get into a short position.

In this sense, the VWAP may be used similarly to moving averages. When the price is above the VWAP line, the market may be interpreted as bullish. At the same time, if it's below the VWAP line, the market may be bearish. This, of course, highly depends on the context of the technical pattern and should be taken with caution.

The VWAP may also be used to identify areas of liquidity. This can be especially useful for institutional traders looking to fill large orders. The indicator helps them determine ideal entry and exit points for large trades, which may decrease their market impact.

VWAP can also be used to measure the efficiency of trade execution. In this sense, buy orders executed below the VWAP may be considered good fills, as they're below the average price of the asset weighted by volume. Conversely, buy orders executed above the VWAP may be considered bad fills, as they're executed above the average price of the asset weighted by volume.

The fact that some large traders buy below the VWAP and sell above it may offer another benefit for the market. These actions push the price closer to the average in both cases. This ensures that large traders don't push prices further from the average with their actions. Bear in mind, whales trade some of the largest sizes, and they could have a substantial impact on the markets otherwise.

VWAP LIMITATIONS

The VWAP is mostly useful as a single-day indicator. Trying to create a VWAP over multiple days could mean that the average is distorted. As such, the VWAP works best for intraday analysis, that is, an analysis that considers one trading day or less.

Like moving averages, the VWAP is a lagging indicator, as it's based on past price data. Similarly, to a moving average, the more data there is, the greater the lag. As such, a 20-minute VWAP will react more quickly to current price movements than a 200-minute VWAP.

It's important to keep in mind that since it's based on past price data, the VWAP doesn't have any predictive qualities.

While the VWAP is a powerful indicator used by many traders, it's not to be interpreted in isolation. For example, we've discussed that an asset may be considered undervalued when the price is below the VWAP line. However, in a strong uptrend, the price may not go below the VWAP for a considerable amount of time.

As such, traders who are waiting for this specific signal might get left on the sideline and miss out on a potential opportunity. With that said, missing out on a trade may not be the end of the world. If a trader's entry strategy dictates a specific event happening, and that event doesn't happen, they shouldn't enter a trade. However, if their strategy is well-thought-out and they're consistently sticking to it, they should be doing well over the long-run. Regardless of the approach, it's crucial to understand and manage the risks.

What is the Parabolic SAR?

The Parabolic SAR is used to determine the direction of the trend and potential reversals. "SAR" stands for *Stop and Reverse*. This refers to the point where a long position should be closed and a short position opened, or vice versa.

The Parabolic SAR appears as a series of dots on a chart, either above or below the price. Generally, if the dots are below the price, it means the price is in an uptrend. In contrast, if the dots are above the price, it means the price is in a downtrend. A reversal occurs when the dots flip to the "other side" of the price.

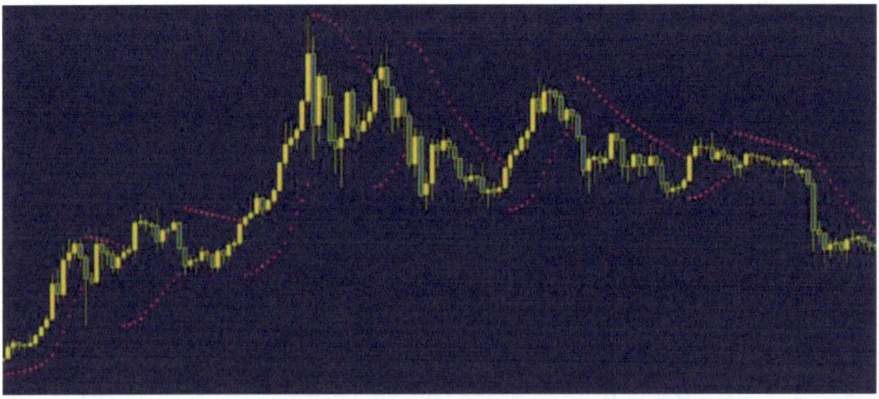

The Parabolic SAR on a Bitcoin chart.

The Parabolic SAR can provide insights into the direction of the market trend. It's also handy for identifying points of trend reversal. Some traders may also use the Parabolic SAR indicator as a basis for their trailing stop-loss. This special-order type moves along with the market and makes sure that investors can protect their profits during a strong uptrend.

The Parabolic SAR is at its best during strong market trends. During periods of consolidation, it may provide a lot of false signals for potential reversals.

What is the Ichimoku Cloud?

The Ichimoku Cloud is a TA indicator that combines many indicators in a single chart. Among the indicators we've discussed, the Ichimoku is certainly one of the most complicated. At first glance, it may be hard to understand its formulas and working mechanisms. But in practice, the Ichimoku Cloud is not as hard to use as it seems, and many traders use it because it can produce very distinct, well-defined trading signals.

As mentioned, the Ichimoku Cloud isn't just an indicator, it's a collection of indicators. It's a collection that provides insights into market momentum, support and resistance levels, and the direction of the trend. It achieves this by calculating five averages and plotting them on a chart. It also produces a "cloud" from these averages which may forecast potential support and resistance areas.

While the averages play an important role, the cloud itself is a key part of the indicator. Generally, if the price is above the cloud, the market may be considered to be in an uptrend. Conversely, if the price is below the cloud, it may be considered to be in a downtrend.

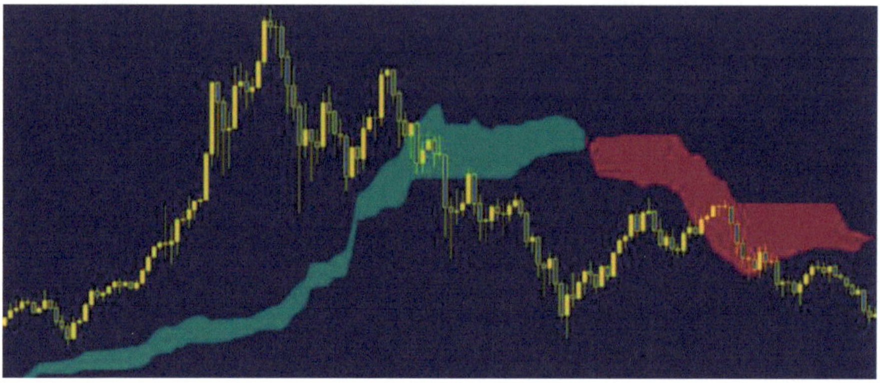

The Ichimoku Cloud on a Bitcoin chart, acting as support, then resistance.

The Ichimoku Cloud may also strengthen other trading signals. The Ichimoku Cloud is difficult to master, but once you get your head around how it works, it can produce great results.

Fundamental analysis vs. Technical analysis – Which is better?

That entirely depends on your trading strategy. Actually, why not use both? Most market analysis methods work best when they're combined with other methods or indicators. This way, there's a bigger chance of finding more reliable investment opportunities. Combining different trading strategies can also help eliminate biases from your decision-making process.

This concept is sometimes referred to as confluence. Confluence traders combine multiple strategies into one that harnesses benefits from all of them. The idea is that the trading opportunities presented by the combined strategies may be stronger than the ones provided by only one strategy.

What drives the financial markets?

The price of an asset is simply determined by the balance of supply and demand. In other words, it's decided by the buyers and sellers. Where supply meets demand, there's a market. But what else can drive the value of a financial asset?

As we've discussed earlier, there can be fundamental factors, such as the state of the economy. In addition, there can be technical factors like the market capitalization of a cryptocurrency. Also, there may be other factors to consider, such as market sentiment or recent news.

However, these are just that – factors to consider. What really determines the price of an asset in a given moment is simply the balance of supply and demand.

What is a market trend?

A market trend is the overall direction where the price of an asset is going. In technical analysis, market trends are typically identified using price action, trend lines, or even key moving averages.

Generally, there are two main types of market trends: bull and bear market. A bull market consists of a sustained uptrend, where prices are continually going up. A bear market consists of a sustained downtrend, where prices are continually going down. In addition, we can also identify consolidating, or "sideways" markets, where there isn't a clear directional trend.

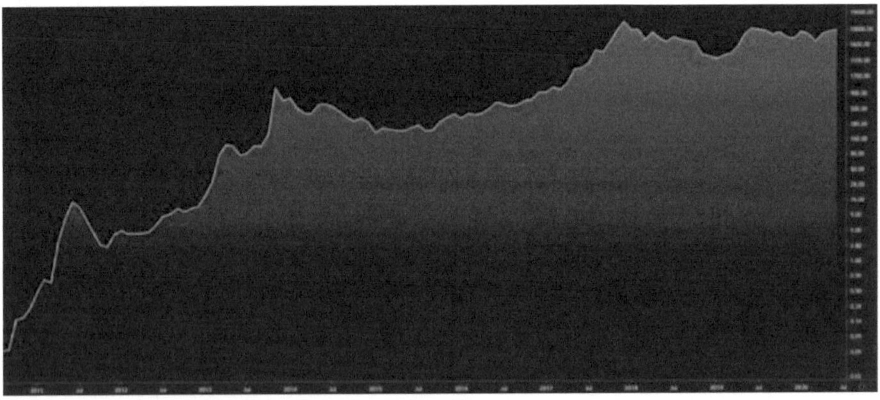

Bitcoin has been in a bull market throughout all its existence.

It's worth noting that a market trend doesn't mean that the price is *always* going in the direction of the trend. A prolonged bull market will have smaller bear trends contained with it, and vice versa. This is simply just the nature of market trends. It's a matter of perspective as it all depends on the time frame you are looking at. Market trends on higher time frames will always have more significance than market trends on lower time frames.

A peculiar thing about market trends is that they can only be determined with absolute certainty in hindsight. You may have heard about the concept of hindsight bias, which refers to the tendency of people to convince themselves that they accurately predicted an event *before it happened*. As you'd imagine, hindsight bias can have a significant impact on the process of identifying market trends and making trading decisions.

What is a market cycle?

You may have heard the phrase that "the market moves in cycles". A cycle is a pattern or trend that emerges at different times. Typically, market cycles on higher time frames are more reliable than market cycles on lower time frames. Even so, you can eventually find small market cycles on an hourly chart just as you may do when looking at decades of data.

Markets are cyclical in nature. Cycles can result in certain asset classes outperforming others. In other segments of the same market cycle, those same asset classes may underperform other types of assets due to the different market conditions.

It's worth noting that it's almost impossible to determine in any given moment *where* we currently are in a market cycle. This analysis can be done with high accuracy only *after* that part of the cycle has concluded. Market cycles also rarely have concrete beginning and endpoints. As it turns out, being in the present moment is an exceptionally biased viewpoint in the financial markets.

What is a Financial Instrument?

In simple terms, a financial instrument is a tradable asset. Examples include cash, precious metals (like gold or silver), a document that confirms ownership of something (like a business or a resource), a right to deliver or receive cash, and many others. Financial instruments can be really complex, but the basic idea is that whatever they are or whatever they represent, they can be traded.

Financial instruments have various types based on different classification methods. One of the classifications is based on whether they are cash instruments or derivative instruments. As the name would suggest, derivative instruments derive their value from something else (like a cryptocurrency). Financial instruments may also be classified as debt-based or equity-based.

But where do cryptocurrencies fall? We could think of them in multiple ways, and they could fit into more than one category. The simplest classification is that they are digital assets. However, the potential of cryptocurrencies lies in building an entirely new financial and economic system.

In this sense, cryptocurrencies form a completely new category of digital assets. What's more, as the ecosystem evolves, many new categories may be established that wouldn't otherwise be possible. Early examples of this can already be seen in the Decentralized Finance (DeFi) space.

What is the spot market?

The spot market is where financial instruments are traded for what's called "immediate delivery". Delivery, in this context, simply means exchanging the financial instrument for cash. This may seem like an unnecessary distinction, but some markets aren't settled in cash instantly. For example, when we're talking about the futures markets, the assets are delivered at a later date (when the futures contract expires).

In simple terms, you could think of a spot market as the place where trades are made "on the spot." Since the trades are settled immediately, the current market price of an asset is often referred to as the *spot price*.

Advantages of Spot Markets

1. Prices are transparent and only rely on supply and demand in the market. This aspect contrasts with the futures market that often contains multiple reference prices. For example, the mark price

in the Binance futures market is derived from other information, including the funding rate, price index, and Moving Average (MA) Basis. In some traditional markets, the mark price might also be affected by interest rates.

2. Spot trading is straightforward to take part in due to its simple rules, rewards, and risks. When you invest $500 on the spot market in BNB, you can calculate your risk easily based on your entry and the current price.

3. You can "set and forget". Unlike derivatives and margin trading, with spot trading, you don't need to worry about being liquidated or getting a margin call. You can enter or exit a trade whenever you want. You also don't need to keep checking your investment, unless you want to make short-term trades.

Disadvantages of Spot Markets

1. Depending on what you're trading, spot markets can leave you with assets that are inconvenient to hold. Commodities are perhaps the best example. If you spot purchase crude oil, you'll have to take physical delivery of the asset. With cryptocurrencies, holding tokens and coins gives you a responsibility to keep them secure and safe. By trading futures derivatives, you can still get exposure to these assets but settle with cash.

2. With certain assets, individuals, and companies, stability is valuable. For example, a company wanting to operate abroad

needs access to foreign currency in the forex market. If they rely on the spot market, expenditure planning and incomes would be very unstable.

3. Potential gains in spot trading are much less than in futures or margin trading. You can leverage the same amount of capital to trade larger positions

What is Margin Trading?

Margin trading is a method of trading using borrowed funds from a third party. In effect, trading on margin amplifies results – both to the upside and the downside. A margin account gives traders more access to capital and eliminates some counterparty risk. How so? Well, traders can trade the same position size but keep less capital on the cryptocurrency exchange.

When it comes to margin trading, you'll often hear the terms **margin** and **leverage**. Margin refers to the amount of capital you commit (i.e., put up from your own pocket). Leverage means the amount that you amplify your margin with. So, if you use 2x leverage, it means that you open a position that's double the amount of your margin. If you use 4x leverage, you open a position that's four times the value of your margin, and so on.

However, be aware of liquidation. The higher leverage you use, the closer the liquidation price is to your entry. If you get liquidated, you'll risk losing your entire margin. So, be very aware of the high risks of trading on margin before getting started

Margin trading is widely used in stock, commodity, and Forex trading, as well as the Bitcoin and cryptocurrency markets. In a more traditional setting, the funds borrowed are provided by an investment broker. When it comes to cryptocurrencies, the funds are typically lent by the exchange in return for a funding fee. In some other cases, however, the borrowed funds may come directly from other traders on the platform. This will usually incur a variable interest rate (funding fee), as the rate is determined by an open marketplace.

What is the derivatives market?

Derivatives are financial assets that base their value on something else. This can be an underlying asset or basket of assets. The most common types are stocks, bonds, commodities, market indexes, or cryptocurrencies.

The derivative product itself is essentially a contract between multiple parties. It gets its price from the underlying asset that's used as the benchmark. Whatever asset is used as this reference point, the core concept is that the derivative product *derives* its value from it. Some common examples of derivatives products are futures contracts, options contracts, and swaps.

According to some estimates, the derivatives market is one of the biggest markets out there. How so? Well, derivatives can exist for virtually any financial product – even derivatives themselves. Yes, derivatives can be created from derivatives. And then, derivatives can be created from *those* derivatives, and so on. Does this sound like a shaky house of cards ready to come crashing down? Well, this may not be so far from the truth. Some argue that the derivatives market played a major part in the 2008 Financial Crisis.

What are forward and futures contracts?

A futures contract is a type of derivatives product that allows traders to speculate on the future price of an asset. It involves an agreement between parties to settle the transaction at a later date called the expiry date. As we've discussed with derivatives, the underlying asset for a contract like this can be any asset. Common examples include cryptocurrency, commodities, stocks, and bonds.

The expiration date of a futures contract is the last day that trading activity is ongoing for that specific contract. At the end of that day, the contract expires to the last traded price. The settlement of the contract is determined beforehand, and it can be either **cash-settled** or **physically-delivered**.

When it's delivered physically, the underlying asset of the contract is directly exchanged. For example, barrels of oil are delivered. When it's settled in cash, the underlying asset isn't exchanged directly, only the value that it represents (in the form of cash or cryptocurrency).

What are perpetual futures contracts?

Futures products are a great way for traders to speculate on the price of an asset. However, what if they want to remain in their position even after the expiry date?

Enter perpetual futures contracts. The main difference between them and a regular futures contract is that they never expire. This way, traders can speculate on the price of the underlying asset without having to worry about expiration.

However, this presents a problem of its own. What if the price of the perpetual futures contract gets really far from the price of the underlying asset? Since there's no expiry date, the perpetual futures market could have a significant, continual disparity with the spot market.

This is why perpetual futures contracts implement a **funding fee** that's paid between traders. Let's imagine that the perpetual futures market is trading higher than the spot market. In this case, the funding rate will be positive, meaning that long positions (buyers) pay the funding fees to short positions (sellers). This encourages buyers to sell, which then causes the price of the contract to drop, moving it closer to the spot price. Conversely, if the perpetual futures market is trading lower than the spot market, the funding rate will be negative. This time, shorts pay longs to incentivize pushing up the price of the contract.

To summarize, if funding is positive, longs pay shorts. If funding is negative, shorts pay longs. Perpetual futures contracts are hugely popular among Bitcoin and cryptocurrency traders.

What are options contracts?

An options contract is a type of derivatives product that gives traders the right, but not the obligation, to buy or sell an asset in the future at a specific price. The main difference between a futures contract and an options contract is that traders are not obligated to settle options contracts.

When traders buy an options contract, they speculate on the price going in a direction.

There are two types of options contracts: **call options** and **put options**. A call option bets on the price going up, while a put option bets on the price going down.

As with other derivatives products, options contracts can be based on a wide variety of financial assets: market indexes, commodities, stocks, cryptocurrencies, and so on.

Options contracts can enable highly complex trading strategies and risk management methods, such as hedging. In the context of cryptocurrencies, options might be the most useful for miners who want to hedge their large cryptocurrency holdings. This way, they're better protected against events that could have a detrimental impact on their funds.

How do options contracts work?

There are two basic types of options, known as puts and calls. Call options give contract owners the right to buy the underlying asset, while put options confer the right to sell. As such, traders usually enter into calls when they expect the price of the underlying asset to increase, and puts when they expect the price to decrease. They may also use calls and puts hoping for prices to remain stable - or even a combination of the two types - to bet in favor or against market volatility.

An options contract consists of at least four components: size, expiration date, strike price, and premium. First, the size of the order refers to the number of contracts to be traded. Second, the expiration date is the date after which a trader can no longer exercise the option. Third, the strike price is the price at which the asset will be bought or sold (in case the contract buyer decides to exercise the option). Finally, the premium is the trading price of the options contract. It indicates the amount an investor must pay to obtain the power of choice. So, buyers acquire contracts from writers (sellers) according to the value of the premium, which is constantly changing, as the expiration date gets closer.

Basically speaking, if the strike price is lower than the market price, the trader can buy the underlying asset at a discount and, after including the premium into the equation, they may choose to exercise the contract to make a profit. But if the strike price is higher than the market price, the holder has no reason to exercise the option, and the contract is deemed useless. When the contract is not exercised, the buyer only loses the premium paid when entering the position.

It is important to note that although the buyers are able to choose between exercising or not their calls and puts, the writers (sellers) are dependent on the buyers' decision. So, if a call option buyer decides to exercise his contract, the seller is obligated to sell the underlying asset. Similarly, if a trader buys a put option and decides to exercise it, the seller is obligated to buy the underlying asset from the contract holder. This means that writers are exposed to higher risks than buyers. While buyers have their losses limited to the premium paid for the contract, writers can lose much more depending on the asset's market price.

Some contracts give traders the right to exercise their option any time before the expiration date. These are usually referred to as American option contracts. In contrast, the European options contracts can only be exercised at the expiration date. It is worth noting, however, that these denominations have nothing to do with their geographical location.

Options Premium

The value of the premium is affected by multiple factors. To simplify, we may assume that the premium of an option is dependent on at least four elements: the underlying asset's price, the strike price, the time left until the expiration date, and the volatility of the corresponding market (or index). These four components present different effects on the premium of calls and put options, as illustrated in the following table.

	Call options premium	Put options premium
Rising asset's price	Increases	Decreases
Higher strike price	Decreases	Increases
Decreasing time	Decreases	Decreases
Volatility	Increases	Increases

Naturally, the asset's price and strike price influence the premium of calls and puts in an opposing way. In contrast, less time usually means lower premium prices for both types of options. The main reason for that is because traders would have a lower probability of those contracts turning in their favor. On the other hand, increased levels of volatility usually cause premium prices to rise. As such, the option contract premium is a result of those and other forces combined.

Options Greeks

Options Greeks are instruments designed to measure some of the multiple factors that affect the price of a contract. Specifically, they are statistical values used to measure the risk of a particular contract based on different underlying variables. Following are some of the primary Greeks and a brief description of what they measure:

- Delta: measures how much the price of an options contract will change in relation to the underlying asset's price. For instance, a Delta of 0.6 suggests that the premium price will likely move $0.60 for every $1 move in the asset's price.
- Gamma: measures the rate of change in Delta over time. So, if Delta changes from 0.6 to 0.45, the option's Gamma would be 0.15.
- Theta: measures price change in relation to a one-day decrease in the contract's time. It suggests how much the premium is expected to change as the options contract gets closer to expiration.
- Vega: measures the rate of change in a contract price in relation to a 1% change in the implied volatility of the

underlying asset. An increase in Vega would normally reflect an increase in the price of both calls and puts.

- Rho: measures expected price change in relation to fluctuations in interest rates. Increased interest rates generally cause an increase in calls and a decrease in puts. As such, the value of Rho is positive for call options and negative for put options.

Hedging

Options contracts are widely used as hedging instruments. A very basic example of a hedging strategy is for traders to buy put options on stocks they already hold. If the overall value is lost in their main holdings due to price declines, exercising the put option can help them mitigate losses.

For example, imagine that Brianna bought 100 shares of a stock at $50, hoping for the market price to increase. However, to hedge against the possibility of stock prices falling, she decided to buy put options with a strike price of $48, paying a $2 premium per share. If the market turns bearish and the stock declines to $35, Brianna can exercise her contract to mitigate losses, selling each share for $48 instead of $35. But if the market turns bullish, she doesn't need to exercise the contract and would only lose the premium paid ($2 per share).

In such a scenario, Brianna would break even at $52 ($50 + $2 per share), while her losses would be limited to -$400 ($200 paid for the premium and $200 more if she sells each share for $48).

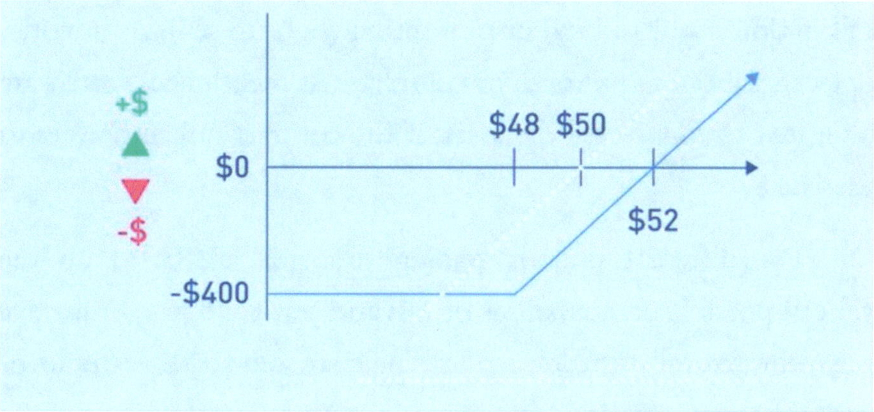

Speculative Trading

Options are also widely used for speculative trading. For instance, a trader who believes that an asset's price is about to go up can buy a call option. If the price of the asset moves above the strike price, the trader can then exercise the option and buy it at a discount. When an asset's price is above or below the strike price in a way that makes the contract profitable, the option is said to be "in-the-Money." Likewise, a contract is said to be "at-the-Money" if on its breakeven point, or "out-of-the-Money" if in a loss.

Basic Strategies

When trading options, traders can employ a wide range of strategies, which are based on four basic positions. As a buyer, one can buy a call option (right to buy) or put option (right to sell). As a writer, one can sell call or put options contracts. As mentioned, writers are obligated to buy or sell the assets if the contract holder decides to exercise it.

The different options trading strategies are based on the various possible combinations of call and put contracts. Protective puts, covered calls, straddle, and strangle are some basic examples of such strategies.

- Protective put: involves buying a put option contract of an asset that is already owned. This is the hedging strategy used by Brianna in the previous example. It is also known as portfolio insurance as it protects the investor from a potential downtrend, while also maintaining their exposure in case the asset's price increases.
- Covered call: involves selling a call option of an asset that is already owned. This strategy is used by investors to generate additional income (options premium) from their holdings. If the contract is not exercised, they earn the premium while keeping their assets. However, if the contract is exercised due to an increase in the market price, they are obligated to sell their positions.

- Straddle: involves buying a call and a put on the same asset with identical strike prices and expiration dates. It allows the trader to profit as long as the asset moves far enough in either direction. Simply put, the trader is betting on market volatility.
- Strangle: involves buying both a call and a put that are "out-of-the-Money" (i.e., strike price above market price for call options and below for put options). Basically, a strangle is like a straddle, but with lower costs for establishing a position. However, a strangle requires a higher level of volatility to be profitable.

Advantages

- Suitable for hedging against market risks.
- More flexibility in speculative trading.
- Allow for several combinations and trading strategies, with unique risk/reward patterns.
- Potential to profit from all the bull, bear, and side-way market trends.
- May be used for reducing costs when entering positions.
- Allow multiple trades to be performed simultaneously.

Disadvantages

- Working mechanisms and premium calculation not always easy to understand.
- Involves high risks, especially for contract writers (sellers)
- More complex trading strategies when compared to conventional alternatives.
- Options markets are often plagued with low levels of liquidity, making them less attractive for most traders.
- The premium value of options contracts is highly volatile and tends to decrease as the expiration date gets closer.

Options vs. Futures

Options and futures contracts are both derivative instruments and, as such, present some common use cases. But despite their similarities, there is a major difference in the settlement mechanism between the two.

Unlike options, futures contracts are always executed when the expiration date is reached, meaning that the contract holders are legally obligated to exchange the underlying asset (or respective value in cash). Options, on the other hand, are only exercised at the discretion of the trader who holds the contract. If the contract holder (buyer) exercises the option, the contract writer (seller) is obligated to trade the underlying asset.

As the name suggests, options give an investor the choice to buy or sell an asset in the future, regardless of the market price. These types of contracts are very versatile and can be used in various scenarios - not only for speculative trading but also for performing hedging strategies.

Yet, it is worth noting that trading options, as well as other derivatives, involves many risks. So before making use of this type of contract, traders should have a good understanding of how it works. It is also important to have a good understanding of the different combinations of calls and puts and the potential risks involved in each strategy. Also, traders should also consider employing risk management strategies along with technical and fundamental analyses to limit the potential losses.

What is the foreign exchange (Forex) market?

The foreign exchange (Forex, FX) market is where traders can exchange one country's currency into another. In essence, the Forex market is what determines the exchange rates for currencies around the world.

We may often think of currencies as "safe haven" assets. Even the term "stablecoin" should imply, in theory, that the asset is somehow safe from volatility. However, while this is true to some extent, currencies can also experience significant market fluctuations. How come? Well, the value of currencies is also determined by supply and demand. In addition, they may also be influenced by inflation or other market forces related to global trade and investment, and geopolitical factors.

How does the Forex market work? Well, currency pairs may be traded by investment banks, central banks, commercial companies, investment firms, hedge funds, and retail Forex traders. The Forex market also enables global currency conversions for international trade settlements.

Forex traders will typically use day trading strategies, such as scalping with leverage, to amplify their returns. We'll cover how exactly that works later in this article.

The Forex market is one of the major building blocks of the modern global economy as we know it. In fact, the Forex market is the largest and most liquid financial market in the world.

What are leveraged tokens?

Leveraged tokens are tradable assets that can give you leveraged exposure to the price of a cryptocurrency without the usual requirements of managing a leveraged position. This means you don't have to worry about margin, collateral, funding, and liquidation.

Leveraged tokens are an innovative financial product that only exist thanks to the power of blockchain. Leveraged tokens were initially introduced by derivatives exchange FTX, but since then have seen various alternative implementations. The main idea behind them, however, is still the same – tokenizing open leveraged positions. What does this mean?

Leveraged tokens represent open perpetual futures positions in a tokenized form. Remember when we discussed how derivatives can be created from derivatives? Leveraged tokens are a prime example since they derive their value from futures positions, which are also derivatives. Leveraged tokens are a great way to get a simple leveraged exposure to a cryptocurrency.

Chapter 9

Increase The Pot

What Is Staking?

You may think of staking as a less resource-intensive alternative to mining. It involves holding funds in a cryptocurrency wallet to support the security and operations of a blockchain network. Simply put, staking is the act of locking cryptocurrencies to receive rewards.

In most cases, you'll be able to stake your coins directly from your crypto wallet, such as Trust Wallet. On the other hand, many exchanges offer staking services to their users. Coinbase Staking lets you earn rewards in an utterly simple way – all you have to do is hold your coins on the exchange. More on this later.

To get a better grasp of what staking is, you'll first need to understand how Proof of Stake (PoS) works. PoS is a consensus mechanism that allows blockchains to operate more energyefficiently while maintaining a decent degree of decentralization (at least, in theory). Let's dive into what PoS is and how staking works.

WHAT IS PROOF OF STAKE (POS)?

If you know how Bitcoin works, you're probably familiar with Proof of Work (PoW). It's the mechanism that allows transactions to be gathered into blocks. Then, these blocks are linked together to create the blockchain. More specifically, miners compete to solve a complex mathematical puzzle, and whoever solves it first gets the right to add the next block to the blockchain.

Proof of Work has proven to be a very robust mechanism to facilitate consensus in a decentralized manner. The problem is, it involves a lot of arbitrary computation. The puzzle the miners are competing to solve serves no purpose other than keeping the network secure. One could argue, this in itself makes this excess of computation justifiable. At this point, you might be wondering: are there other ways to maintain decentralized consensus without the high computational cost?

Enter Proof of Stake. The main idea is that participants can lock coins (their "stake"), and at particular intervals, the protocol randomly assigns the right to one of them to validate the next block. Typically, the probability of being chosen is proportional to the number of coins – the more coins locked up, the higher the chances.

This way, what determines which participants create a block isn't based on their ability to solve hash challenges as it is with Proof of Work. Instead, it's determined by how many staking coins they are holding.

Some might argue that the production of blocks through staking enables a higher degree of scalability for blockchains. This is one of the reasons the Ethereum network is planned to migrate from PoW to PoS in a set of technical upgrades collectively referred to as ETH 2.0.

One of the early appearances of Proof of Stake may be attributed to Sunny King and Scott Nadal in their 2012 paper for Peercoin. They describe it as a *"peer-to-peer cryptocurrency design derived from Satoshi Nakamoto's Bitcoin."*

The Peercoin network was launched with a hybrid PoW/PoS mechanism, where PoW was mainly used to mint the initial supply. However, it wasn't required for the long-term sustainability of the network, and its significance was gradually reduced. In fact, most of the network's security relied on PoS.

WHAT IS DELEGATED PROOF OF STAKE (DPoS)?

An alternative version of this mechanism was developed in 2014 by Daniel Larimer called Delegated Proof of Stake (DPoS). It was first used as a part of the BitShares blockchain, but soon after, other networks adopted the model. These include Steem and EOS, which were also created by Larimer.

DPoS allows users to commit their coin balances as votes, where voting power is proportional to the number of coins held. These votes are then used to elect a number of delegates who manage the blockchain on behalf of their voters, ensuring security and consensus. Typically, the staking rewards are distributed to these elected delegates, who then distribute part of the rewards to their electors proportionally to their individual contributions.

The DPoS model allows for consensus to be achieved with a lower number of validating nodes. As such, it tends to enhance network performance. On the other hand, it may also result in a lower degree of decentralization as the network relies on a small, select group of validating nodes. These validating nodes handle the operations and overall governance of the blockchain. They participate in the processes of reaching consensus and defining key governance parameters.

Simply put, DPoS allows users to signal their influence through other participants of the network.

How does staking work?

As we've discussed before, Proof of Work blockchains rely on mining to add new blocks to the blockchain. In contrast, Proof of Stake chains produce and validate new blocks through the process of staking. Staking involves validators who lock up their coins so they can be randomly selected by the protocol at specific intervals to create a block. Usually, participants that stake larger amounts have a higher chance of being chosen as the next block validator.

This allows for blocks to be produced without relying on specialized mining hardware, such as ASICs. While ASIC mining requires a significant investment in hardware, staking requires a direct investment in the cryptocurrency itself. So, instead of competing for the next block with computational work, PoS validators are selected based on the number of coins they are staking. The "stake" (the coin holding) is what incentivizes validators to maintain network security. If they fail to do that, their entire stake might be at risk

While each Proof of Stake blockchain has its particular staking currency, some networks adopt a two-token system where the rewards are paid in a second token.

On a very practical level, staking just means keeping funds in a suitable wallet. This enables essentially anyone to perform various network functions in return for staking rewards. It may also include adding funds to a staking pool, which we'll cover shortly.

How are staking rewards calculated?

There's no short answer here. Each blockchain network may use a different way of calculating staking rewards.

Some are adjusted on a block-by-block basis, taking into account many different factors. These can include:

- how many coins the validator is staking
- how long the validator has been actively staking
- how many coins are staked on the network in total?
- the inflation rates
- other factors

For some other networks, staking rewards are determined as a fixed percentage. These rewards are distributed to validators as a sort of compensation for inflation. Inflation encourages users to spend their coins instead of holding them, which may increase their usage as cryptocurrency. But with this model, validators can calculate exactly what staking reward they can expect.

A predictable reward schedule rather than a probabilistic chance of receiving a block reward may look favorable to some. And since this is public information, it might incentivize more participants to get involved in staking.

WHAT IS A STAKING POOL?

A staking pool is a group of coin holders merging their resources to increase their chances of validating blocks and receiving rewards. They combine their staking power and share the rewards proportionally to their contributions to the pool.

Setting up and maintaining a staking pool often requires a lot of time and expertise. Staking pools tend to be the most effective on networks where the barrier of entry (technical or financial) is relatively high. As such, many pool providers charge a fee from the staking rewards that are distributed to participants.

Other than that, pools may provide additional flexibility for individual stakers. Typically, the stake has to be locked for a fixed period and usually has a withdrawal or unbinding time set by the protocol. What's more, there's almost certainly a substantial minimum balance required to stake to disincentivize malicious behavior.

Most staking pools require a low minimum balance and append no additional withdrawal times. As such, joining a staking pool instead of staking solo might be ideal for newer users.

What is Cold Staking?

Cold staking refers to the process of staking on a wallet that's not connected to the Internet. This may be done using a hardware wallet, but it's also possible with an air-gapped software wallet.

Networks that support cold staking allow users to stake while securely holding their funds offline. It's worth noting that if the stakeholder moves their coins out of cold storage, they'll stop receiving rewards.

Cold staking is particularly useful for large stakeholders who want to ensure maximum protection of their funds while supporting the network.

What is Crypto Lending?

Crypto lending refers to a type of Decentralized Finance that allows investors to lend their cryptocurrencies to different borrowers. This way, they will get interest payments in exchange, also called "crypto dividends". Many platforms that specialize in lending crypto also accept stablecoins, on top of cryptos.

One of the benefits of investing in cryptocurrency is the ability to use your crypto holdings as collateral for a loan, even if your holdings are relatively small. In traditional markets, a similar practice is called securities-based lending, but it is typically off-limits to all but high-net worth clients of private banks and large financial institutions.

Crypto loans are much more accessible, and they offer "hodl" investors a way to achieve liquidity from their investment without selling it. In this article, we'll discuss the advantages, risks, and tax implications of bitcoin and crypto loans, as well as introduce some of the best crypto lending platforms.

Custodial crypto (CeFi) loans

Centralized finance (CEFi) loans are custodial; a central entity takes custody of collateral. In this situation, a trader cannot access his or her collateralized assets; the lender controls the assets' private keys.

The takeaway here is that although custodial crypto loans are still far more accessible and affordable than traditional loans, they still depend on a centralized lending provider to enforce their terms. Around 80 percent of crypto loans are currently custodial, but this ratio is changing quickly.

Non-custodial (DeFI) crypto loans

Decentralized finance (DeFi) loans are non-custodial. Rather than depending on a central organization to enforce the terms of the loan, they depend on smart contracts. If a trader takes out a DeFi crypto loan, the trader retains control of their assets' keys—unless they default on the loan.

DeFi platforms cannot directly lend fiat currency; traders receive stablecoins that can then be exchanged for cash. DeFi loans tend to have a higher interest rate than custodial loans.

What do I need to take out a crypto loan?

Compared to the process of applying for a traditional loan, applying for a crypto loan requires relatively little. Credit checks are typically not required; rather, the amount of the loan you will be approved for depends upon the amount of collateral you are able to use. The loan-to-value (LTV) ratio is the ratio between the amount of the loan and the value of the collateral. If you put up $10,000 worth of crypto as collateral and receive a $6,000 loan, your loan's LTV ratio is 60 percent. Because crypto markets are volatile, LTV ratios on crypto loans are typically low.

Leading DeFi lending platforms

Oasis Borrow (MakerDAO)

Oasis Borrow is the hub for crypto loans from the Maker ecosystem, which supports the DAI stablecoin. Maker loans are appealing because they are essentially loans to yourself. Anyone with crypto and a MetaMask wallet can open a Maker Vault and begin minting DAI. With Oasis Borrow, you put up an asset as collateral and agree to either mint a certain amount of DAI or pay back your loan before unlocking your asset. So, essentially you are borrowing crypto that your collateral asset will be used to generate.

Maker offers both a 50% and 75% LTV rate for crypto loans, with the higher rate also receiving higher stability fees (which are variable). The platform's smart contracts automatically liquidate accounts whose LTVs fall below the agreed upon rate, which also incurs a liquidation penalty of 13% or more.

Currently, Maker is the largest DeFi lending project, with more than $7.3B in locked assets and more than $2.7B in outstanding loans.

Compound Finance

Compound is another large DeFi platform. Its users supply cryptocurrency to support the protocol and earn interest on their deposit. The amount they've supplied is represented by cTokens, whose value increases with time. However, this cToken supply balance can also be used as collateral for a loan.

Compound's LTV rates are typically slightly higher than MakerDAO's. However, Compound not only has no borrowing fees, but it also has a lower liquidation threshold and will only liquidate 50% of an under-collateralized loan, with a liquidation penalty fixed at 8%.

Aave

Aave is an innovative protocol with a wider range of loan options than other large protocols. Like Compound, borrowers put up collateral that is used to support the protocol; their contribution is represented in aTokens.

However, beyond regular DeFi loans, AAVE offers flash loans, which it pioneered, and fixed interest rate loans. It boasts a higher LTV rate than other large competitors and very low borrowing rates. Additionally, it accepts 24 coins as collateral, compared to Maker's 18 and Compound's 9.

Aave is the highest valued DeFi lending project because of its perceived dynamism and innovation.

Alchemix

Alchemix is a smaller DeFi loan platform that uses a new but intriguing method to provide loans that "pay themselves back over time." In very simple terms, it works like this: users deposit DAI into a smart contract and in exchange receive a token that represents the deposit's future yield farming potential. This token is called alUSD and can be transmuted 1-1 for DAI on the Alchemix platform or traded on a DeFi exchange like Sushiswap.

The smart contract then transfers the deposited assets into a Yearn vault that mints DAI. As the yield harvests, Alchemix users' alUSD debt decreases, meaning that ultimately the deposit's yield harvest automatically "pays back" the loan.

Alchemix users can mint an alUSD coin worth up to half of the asset they deposit. If we translated this ratio into more typical crypto loan terms, Alchemix offers an LTV of 50%. Although it currently only accepts DAI deposits, it expects to accept more stablecoins soon.

Leading CeFi lending platforms

Celsius Network

Celsius Network is a popular loan platform, with over $10B in assets and 485,000 users. This popularity is in part thanks to its extremely low (for now) borrowing rates, which start at just 1% (average CeFi borrowing rates at the time of writing are around 4%). Celsius supports 25 coins and offers flexible LTV rates, although they are capped at 50%.

Nexo

Another popular CeFi loan platform, Nexo is differentiated because it offers $375 million of insurance on all custodial assets. The platform has $12B in assets and more than 1.5 million users.

Nexo typically has slightly higher LTV rates and slightly lower borrowing rates. It supports 18 currencies.

Unchained Capital

Among CeFi lenders, Unchained Capital stands out because of its multisig collaborative custody model, which gives borrowers more transparency into their assets and increases security. In this system, accessing collateralized assets requires three private keys. One is controlled by the borrower, one by Unchained Capital, and one by a third-party key agent.

Unchained Capital only offers bitcoin loans and only lends in the United States. Additionally, to use the platform, borrowers must use a hardware wallet from Trezor, Ledger, or Coldcard. Its higher security comes with a somewhat higher barrier to borrowing: it has lower LTV rates and higher interest rates than most CeFi providers.

BlockFi

BlockFi is a huge CeFi player, based and regulated in the U.S. and backed by large financial institutions such as Valar Ventures, Winklevoss Capital, Galaxy Digital, Susquehanna, Akuna Capital, and Fidelity. Motley Fool describes BlockFi as a good fit for beginner and intermediate investors looking to bridge the gap between traditional finance and crypto.

Chapter 10

MINT (Management of Innovation and New Technology)

A Quickbook to Crypto Collectibles and Non-fungible Tokens (NFTs)

Bitcoin introduced the concept digital scarcity. Previously, the cost of digitally copying something was next to nothing. With blockchain technology, creators have the ability to map the digital to the real world.

Non-fungible tokens (NFTs), also known as crypto collectibles, are unique & limited in quantity.

They serve as a building block in our digital economy. Businesses have already experimented with NFTs including gaming, digital identity, licensing, certificates, and fine art. NFTs allow for fractional ownership of high-value items.

WHAT IS A NON-FUNGIBLE TOKEN?

A non-fungible token (NFT) is a cryptographic token on a blockchain. It's an asset that's either digital or tokenized versions of real-world assets. NFTs are not interchangeable with each other. Rather they serve as proof of authenticity and ownership in a virtual network.

Fungibility means that an asset's individual units are interchangeable. For example, fiat (cash) currencies are fungible because dollars are interchangeable with coins. A ten-dollar bill is interchangeable with any other genuine ten-dollar bill.

Fungibility is appealing because it enables free exchange. Hence there's no way to know the history of each dollar bill. This is not a beneficial trait for collectible items because people value the story behind digital assets.

Imagine a digital asset similar to Bitcoin but instead each coin is assigned a unique identifier. This would make each coin different from all the other units (i.e., non-fungible). Essentially, this is what an NFT is.

How do NFTs work?

There are several platforms to design & sell NFT's. Currently the most popular is ERC-721. It's the standard for trading non-fungible assets on the **Ethereum** blockchain.

A more recent, improved standard is ERC-1155. It benefits users by offering a single contract with fungible and non-fungible tokens. This opens a range of creative possibilities. The standard issue of NFTs fosters interoperability. This means assets can be easily transferred between different applications with relative ease.

Trust Wallet archives your NFT collection. Similar to a token, your NFT will exist on a specific address. This password cannot be replicated or transferred without the owner's permission.

NFTs can be traded in open or closed marketplaces. Public open marketplaces include Treasureland, BakerySwap, and Juggerworld on BSC, and OpenSea on Ethereum. Here markets connect buyers & sellers, offering products with uniquely valued tokens.

NFT price changes based on market supply and demand.

How do NFT's have value? Truly, value is a shared belief. Gucci's most recent NFT album was purchased for millions of dollars. Digital collectibles are valued by the audiences they attract. The key to leveraging an NFT is becoming aware of cultural trends & patterns to predict when the market moves in favor of our personal collection.

WHAT CAN NFTS BE USED FOR?

NFTs can be used by apps to issue unique digital items and crypto-collectibles. These decentralized applications (DApps) promote tokens as collectibles or an investment item. In the gaming many online worlds already had their own marketplace: Think of Microsoft cards for Xbox games or credit on an IPhone game. The blockchain serves to tokenize gaming assets to mitigate the virtual inflation.

In the real-world, NFTs can represent fractions of real-world assets. A piece of a tangible asset, NFT's can then be stored and traded as tokens on a blockchain. This concept introduces liquidity to markets that otherwise wouldn't have much, such as fine art, real estate & rare collectible items. In the real world, depreciation affects the value of an item. Real estate must be maintained, fine art requires refurbishing & collectibles may lose value over time. The metaverse allows for a fraction of these assets to be digitally held, maintained or sold at a value at or above its purchased price/

NFTs promote the growth of a strong digital footprint & identity. Individual ownership data on the blockchain increases privacy and coincidingly data integrity: Virtual assets are difficult to steal or lose. In correlation, easy transfers have the potential to reduce friction in the global economy.

NFT Projects You Should Know

Public interest in NFT's is astronomical in 2021. The current NFT bubble has introduced us to decentralization, virtual marketplaces & digital assets. The value proposition of NFTs is admittedly higher for people not involved with blockchain technology. This chapter discusses mainstream NFT's as you know of them today. As we've discussed, NFTs are unique digital items secured by cryptography. They represent digital collectibles, game items, pets, & memes.

Beeple NFTs

The Beeple NFT is one of the rarest in the crypto universe. It was created by artist Mike Winkelmann, better known as Beeple. He sold an NFT for $69M worth of ETH by auction house Christie's. Never selling a single piece of print art, he quickly became one of the highest-valued artists in the world.

What makes his work so valuable? It began with his large fanbase. Beeple fans follow his series where he creates and uploads artwork every day. Never missing a single day in the last 14 years, his most expensive piece, titled "Everydays - The First 5000 Days", is a collage of his artwork from the first 5000 days of the project.

NBA Top Shot

NBA Top Shot is one the most successful NFT projects. This NFT project has the most users in any case. NBA Top Shot is a project by Dapper Labs, users purchase packs of unique digital artwork that feature famous NBA moments as NFTs. This NBA project doesn't feel like a blockchain product, which lowers the barrier of entry for most people.

The project avoids speculation by stripping away from decentralization to improve user experience. IN NBA Top Shot, many users don't know they're using a blockchain-based product.

How to Make Your Own NFTs

Start by researching the Opensea NFT Marketplace, the Mintable or DeFi platforms. We prefer BSC based on its low fees, quick transaction times, and has a growing NFT community.

How's it done?

1. Users need to fill in personal NFT details & a product description.
2. Creators than upload a digital art or file.
3. Lastly, artists must pay a minting fee.
4. Interested in selling? Post it on a NFT marketplace.

Celebrities, digital artists, and creatives are all interested in publishing artwork on the blockchain as a non-fungible token. The possibilities are endless when minting NFTs. NFT minting proves authenticity and ownership of creative works.

To mint an NFT artists must have the following 3 things:

1. A song, artwork, or collectible
2. Some crypto to pay the minting fee
3. A cryptocurrency wallet to store your crypto

Users must then pick a blockchain used to create a non-fungible token.

How to mint an NFT on BakerySwap

You can craft NFTs within 5 minutes on BakerySwap. To do this, have BNB ready for any minting fees: (0.01 BNB) & a crypto wallet to store the BNB. Minting fees change based on the price of BNB. If you don't have a wallet, review Trust Wallet for mobile access and MetaMask for desktop. With BakerySwap, you can choose to mint pictures, GIFs, videos or audio.

1. Go to the BakerySwap NFT Marketplace and connect your wallet.
2. Once connected, click [Mint Artworks] to start creating your NFT.

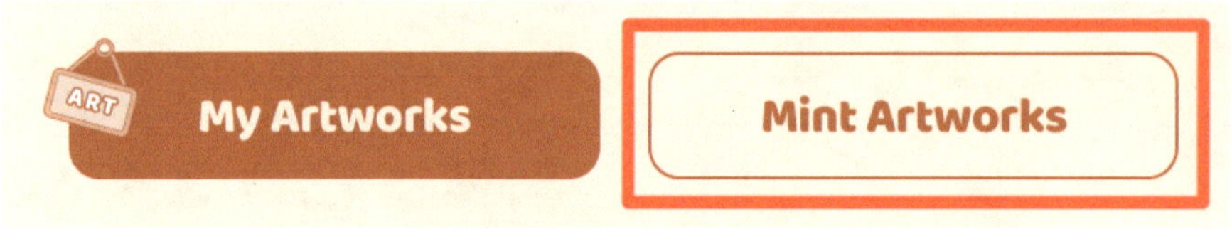

3. Fill out the required fields with all the collectible info.

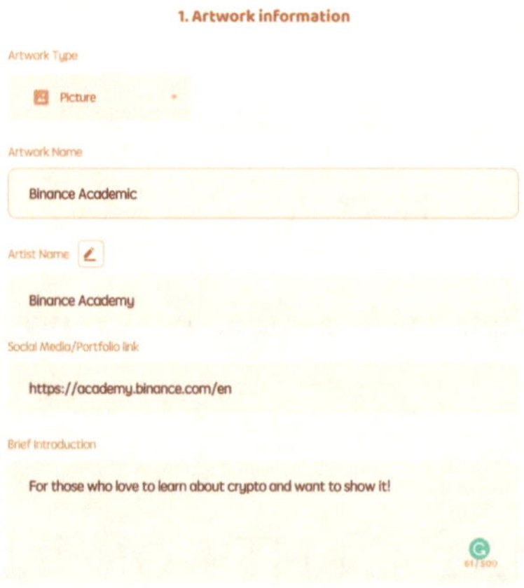

4. Underneath the NFT details, click the [+] icon to upload a file. Here you can see the current minting fee underneath the anti-plagiarism declaration.

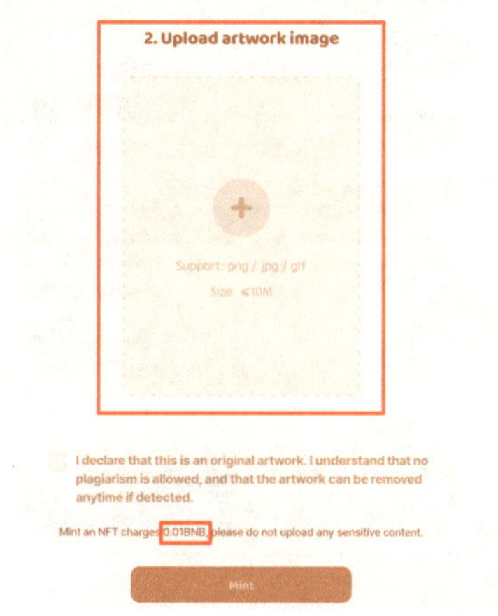

5. Agree to the anti-plagiarism declaration before clicking [Mint].

6. If using MetaMask, a window will pop up confirming payment for your minting fee.

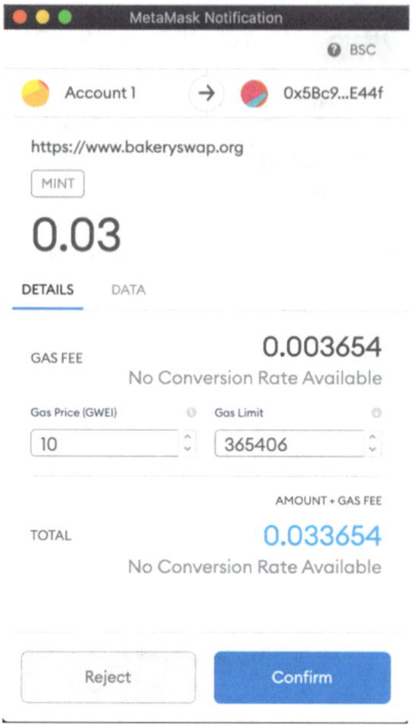

7. View your NFT on the blockchain with BscScan. It's now available to transfer and trade on the blockchain.

How to mint an NFT on Treasureland

Treasureland is an NFT marketplace linked to Dego Finance, a decentralized finance (DeFi) protocol and ecosystem. Minting NFTs with Treasureland is free for the creator & anyone who purchases the NFT pays a fee. Treasureland allows users to mint pictures and GIFs, choosing between a single NFT, a series, or a set of creator royalties.

1. Go to the Treasureland Create section & connect a crypto wallet.
2. On the NFT creation page, fill in all NFT details and upload a image or gif.

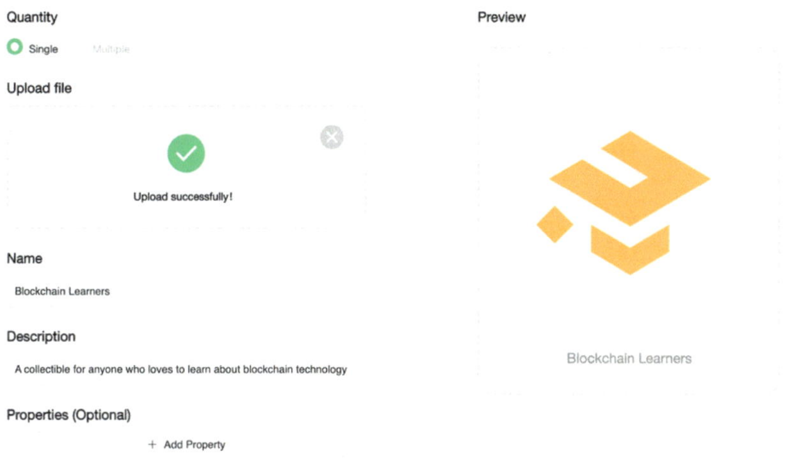

3. Confirm the Treasureland locked royalties at 10% before creating a NFT.

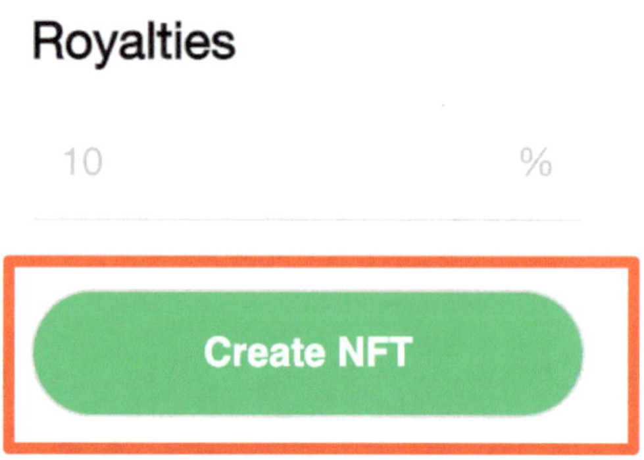

4. View your NFT or post it for sale. If you want to transfer it, you'll have to pay a minting fee.

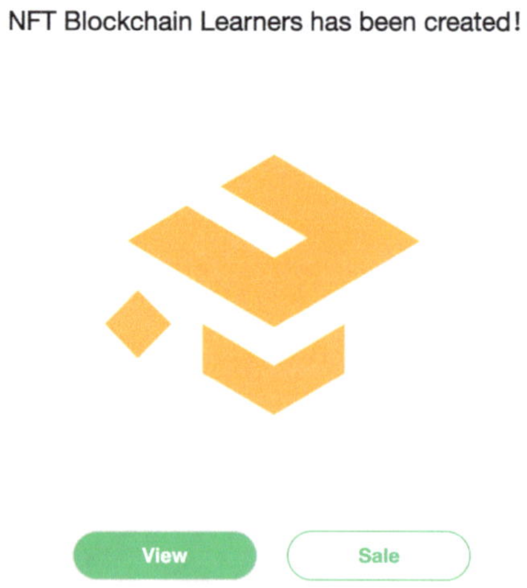

HOW DO I SEND MY NFT TO SOMEONE ELSE?

Once you've purchased or created your custom NFT, you can send it to someone else directly from a wallet. If the wallet has an NFT section, pick which NFT you'd like to send & hit confirm. (Trust Wallet & MetaMask).

The wallet receiving the NFT must have an accurate deposit address for the token you've shared. On the Ethereum network with an ERC-721 NFT, recipients would enter a Ethereum ERC-721 deposit address from their wallet.

SHARE NFT OWNERSHIP AND SALE PROCEEDS WITH OTHERS

When creating group projects with NFTs, its tricky to share partnered commissions. For common NFT forms (ERC721, ERC1155, BEP721, BEP1155), assets can only have 1 owner. In the future we can expect for the proceeds of a first sale to be split among multiple wallets when applicable.

This feature is not built into a specific token's code. Instead, it relies on the marketplace artists use. Not currently available, this online marketplace feature may be offered as soon as 2022. We recommend researching which credible, verified exchanges allow the NFT's sale amount to be shared. For 100% control over personal

NFTs, should store them in a non-custodial wallet on a decentralized platform.

Chapter 11

To Infinity and Beyond

What Is the Metaverse?

First, let's take the word Meta and its etymology. The ancient Greek civilization coined this word to mean 'beyond' or above. Now, let's look at the word verse. Its original text implies it to mean "To turn" in the manner a farmer would his plow. Interesting enough, we get a turn that is beyond physical matter. Nevertheless, the metaverse is a 3D universe virtual space accessible only via an online network: Imagine an evolved civilization that exists solely on an invisible plane. In the metaverse, people can work, meet, grow, and socialize. The metaverse isn't fully in existence: no one can live a sustainable digital life with no real-world interactions. However, some platforms have metaverse- elements. Video games & online concerts currently provide the closest metaverse experience on a public level. Developers are expanding the boundaries of gaming experience with digital events and virtual economies. Besides supporting gaming or social media, the metaverse will combine economies, digital identity, decentralized governance, and other applications. Even today, user creation and ownership of valuable items and currencies help develop a single, united metaverse. All these features provide blockchain the potential to power this future technology.

Not required to join, cryptocurrencies were designed to exist metaverse. Compare them to the currencies available to us now such as the Euro or Yen. Cryptocurrencies promote a digital economy with a range of utility tokens & virtual collectibles (NFTs). Held in crypto wallets, such as Trust Wallet or MetaMask. Also, these ecosystems would be reliable governed by transparent blockchain systems.

These applications already exist, providing participants with a livable income. How is this possible? Modern experts now have the ability to convert in game rewards into capital by using a crypto waller. A mix of the blockchain world & virtual reality apps, tech companies are already tapping it's the Metaverse potential. In 2021, FaceBook formally changed their business name to Meta. Do you understand why?

The barriers between the financial, virtual, and physical have continued to fall. The technology available to us now has defined the capacity of the 21st century. These privately owned devices are relied on everyday & their networks cultivate the world our families will inherit in the future. The crypto ecosystem has introduced NFTs, blockchain games, and crypto payments as part of a worldwide metaverse.

WHY ARE VIDEO GAMES LINKED TO THE METAVERSE?

With such a high interest in 3D virtual reality, video games offer the most succinct metaverse experience. Consider, Roblox, for example, where users can host virtual events such as concerts and mock meetings. Players participate in the "cyberspace" for no reason other than personal interest. For instance, in the game Fortnite, 12.3 million players recently attended Travis Scott's virtual in-game music tour.

HOW DOES CRYPTO FIT INTO THE METAVERSE?

Gaming provides the 3D aspect of the metaverse but doesn't cover everything needed in a virtual world that can cover all aspects of life. Crypto can offer the other key parts required, such as digital proof of ownership, transfer of value, governance, and accessibility. But what do these mean exactly?

If, in the future, we work, socialize, and even purchase virtual items in the metaverse, we need a secure way of showing ownership. We also need to feel safe transferring these items and money around the metaverse. Finally, we will also want to play a role in the decision-making taking place in the metaverse if it will be such a large part of our lives.

Some video games contain some basic solutions already, but many developers use crypto and blockchain instead as a better option. Blockchain provides a decentralized and transparent way of dealing with the topics, while video-game development is more centralized.

Blockchain developers also take influence from the video game world too. Gamification is common in Decentralized Finance (DeFi) and GameFi. It seems there will be enough similarities in the future that the two worlds may become even more integrated. The key aspects of blockchain suited to the metaverse are:

1. **Digital proof of ownership:** By owning a wallet with access to your private keys, you can instantly prove ownership of activity or an asset on the blockchain. For example, you could show an exact transcript of your transactions on the blockchain while at work to show accountability. A wallet is one of the most secure and robust methods for establishing a digital identity and proof of ownership.
2. **Digital collectibility:** Just as we can establish who owns something, we can also show that an item is original and unique.

For a metaverse looking to incorporate more real-life activities, this is important. Through NFTs, we can create objects that are 100% unique and can never be copied exactly or forged. A blockchain can also represent ownership of physical items.

3. **Transfer of value**: A metaverse will need a way to transfer value securely that users trust. In-game currencies in multiplayer games are less secure than crypto on a blockchain. If users spend large amounts of time in the metaverse and even earn money there, they will need a reliable currency.

4. **Governance:** The ability to control the rules of your interaction with the metaverse should also be important for users. In real life, we can have voting rights in companies and elect leaders and governments. The metaverse will also need ways to implement fair governance, and blockchain is already a proven way of doing this.

5. **Accessibility:** Creating a wallet is open to anyone around the world on public blockchains. Unlike a bank account, you don't need to pay any money or provide any details. This makes it one of the most accessible ways to manage finances and an online, digital identity.

6. **Interoperability:** Blockchain technology is continuously improving compatibility between different platforms. Projects like Polkadot (DOT) and Avalanche (AVAX) allow for creating custom blockchains that can interact with each other. A single

metaverse will need to connect multiple projects, and blockchain technology already has solutions for this.

WHAT IS A METAVERSE JOB?

As we mentioned, the metaverse will combine all aspects of life in one place. While many people already work at home, in the metaverse, you will be able to enter a 3D office and interact with your colleagues' avatars. Your job may also be metaverse related and provide you with income directly usable in the metaverse. In fact, these kinds of jobs already exist in a similar form.

GameFi and play-to-earn models now provide steady income streams for people worldwide. These online jobs are great candidates for metaverse implementation in the future, as they show that people are willing to spend their time living and earning in virtual worlds. Play-to-earn games like Axie Infinity and Gods Unchained don't even have 3D worlds or avatars. However, it's the principle that they could be part of the metaverse as a way to earn money entirely in the online world.

METAVERSE EXAMPLES

While we don't yet have a single, linked metaverse, we have plenty of platforms and projects similar to the metaverse. Typically, these also incorporate NFTs and other blockchain elements. Let's look at three examples:

SecondLive

SecondLive is a 3D virtual environment where users control avatars for socializing, learning, and business. The project also has an NFT marketplace for swapping collectibles. In September 2020, SecondLive hosted Binance Smart Chain's Harvest Festival as part of its first anniversary. The virtual expo showcased different projects in the BSC ecosystem for users to explore and interact with.

Axie Infinity

Axie Infinity is a play-to-earn game that's provided players in developing countries an opportunity to earn consistent income. By purchasing or being gifted three creatures known as Axies, a player can start farming the Smooth Love Potion (SLP) token. When sold on the open market, someone could make roughly $200 to $1000 (USD) depending on how much they play and the market price.

While Axie Infinity doesn't provide a singular 3D character or avatar, it gives users the opportunity for a metaverse-like job. You might have already heard the famous story of Filipinos using it as an alternative to full-time employment or welfare.

Decentraland

Decentraland is an online, digital world that combines social elements with cryptocurrencies, NFTs, and virtual real estate. On top of this, players also take an active role in the governance of the platform. Like other blockchain games, NFTs are used to represent cosmetic collectibles. They're also used for LAND, 16x16 meter land parcels that users can purchase in the game with the cryptocurrency MANA. The combination of all of these creates a complex crypto-economy.

WHAT'S THE FUTURE OF THE METAVERSE?

Facebook is one of the loudest voices for the creation of a unified metaverse. This is particularly interesting for a crypto-powered metaverse due to Facebook's Diem stablecoin project. Mark Zuckerberg has explicitly mentioned his plans to use a metaverse project to support remote work and improve financial opportunities for people in developing countries. Facebook's ownership of social media, communication, and crypto platforms give it a good start combining all these worlds into one. Other large tech companies are also targeting the creation of a metaverse, including Microsoft, Apple, and Google.

When it comes to a crypto-powered metaverse, further integration between NFT marketplaces and 3D virtual universes seems like the next step. NFT holders can already sell their goods from multiple sources on marketplaces like OpenSea and BakerySwap, but there isn't yet a popular 3D platform for this. At a bigger scale, blockchain developers might develop popular metaverse-like applications with more organic users than a large tech giant.

While a single, united metaverse is likely a long way off, we already can see developments that may lead to its creation. It looks to be yet another sci-fi use case for blockchain technology and cryptocurrencies. If we will ever really reach the point of a metaverse is unsure. But in the meantime, we can already experience metaverse-like projects and continue to integrate blockchain more into our daily lives.

Chapter 12

Secure The Bag

Secure Your Cryptocurrency

You should always keep your crypto secure whether you're purchasing, storing, or investing. Losing your coins and tokens is, in the vast majority of cases, permanent.

If you trade cryptocurrencies on centralized exchanges, use ones that are regulatory compliant with KYC and AML checks. Peer-to-peer trading and decentralized exchanges with audits have the best chance of security.

There are multiple options when it comes to storing your crypto securely. You can keep your crypto on a regulated exchange, which is practical for newcomers and traders. However, you don't own the keys to the wallet.

A non-custodial wallet where you own the keys provides more security, and the more secure option is to keep it in a wallet not connected to the internet like a cold storage device. In both cases, keep your private keys safe in an offline, secure place.

Use audited DApps to improve your security and regularly check which DApps have permission to use your wallet. Remove these permissions as soon as you're finished using the DApp.

At the core of cryptocurrencies is the idea of self-sovereignty – the notion that a user can act as their own bank. Secure your funds properly, and they'll be harder to reach than even the most well-guarded of bank vaults. Fail to do so, and you run the risk of someone remotely emptying your digital wallet.

Learning to secure your digital coins properly is a vital step as you journey down the cryptocurrency rabbit hole. It's not just all about storage, either. Nowadays, many cryptocurrency holders interact with DApps in the DeFi world, so you should also learn how to use your coins securely.

Just like you wouldn't allow an untrustworthy business to handle your money, you also shouldn't trust your coins with any random DApp. The same goes for exchanges where you purchase and trade crypto. In this guide, we'll discuss some of the best techniques for keeping your crypto assets safe wherever they are.

PURCHASING CRYPTO SECURELY

There are many places where you can purchase cryptocurrencies nowadays. The list includes centralized exchanges, decentralized exchanges (DEX), crypto-ATMs, peer-to-peer options, and more. Not every choice offers the same amount of security, and each has its advantages and disadvantages. For most users, using reputable, centralized exchanges provides the best mix of ease-of-use and security.

PICKING A SECURE EXCHANGE

For a centralized exchange like Coinbase, increasing regulation, Anti-Money Laundering (AML) measures, and Know Your Customer (KYC) checks provide security. While exchanges in the early days of crypto had their issues, governments and exchange operators have since improved the situation significantly.

To use an exchange, you'll need to transfer your funds into its custodial wallet. Giving the exchange responsibility for your coins can provide some security depending on your outlook. If you aren't familiar with wallets or are new to cryptocurrencies, you may be more secure using the exchange's wallet. This saves you from accidentally locking yourself out of your wallet and losing your crypto.

However, some people prefer the security of controlling their funds directly. You may have already heard the phrase "not your keys, not your coins". If you don't actually own the wallet, then someone else can control your crypto. You can check our storage section later on for more information.

If you've decided on using a peer-to-peer service or a decentralized exchange, there are a few signs to look for to improve your security. With a DEX, check for an audit from a reputable source.

If you need to use a peer-to-peer service, make sure it requires KYC for both buyers and sellers. Ideally, it should also offer an escrow service. While it doesn't remove the risks completely, a third party holding your funds in escrow provides both the buyer and seller more protection from scams.

HOW TO SECURE YOUR ACCOUNT

If you signed up for your exchange or chosen trading method, follow standard good practices to keep your account safe. These tips are no different from those you would use for your online bank account or other sensitive information. Preventing people from getting access to your account and its funds is easy by:

1. **Using a strong password you regularly change.** The password shouldn't include identifiable personal information like your date of birth, for example. Make sure it's also long, is unique to that account, and contains symbols, numbers, and lowercase and uppercase letters.

2. **Enabling Two-Factor Authentication (2FA).** If your password is compromised, 2FA using your mobile device, authenticator app, or YubiKey acts as a second level of protection. You need to use both your password and the 2FA method together when logging in.
3. **Watching out for phishing attacks and scams via email, social media, and private messages.** Fraudsters frequently impersonate exchanges and trusted individuals to try and steal your funds. You also shouldn't download software from unknown sources as it may contain malware.

HOW TO STORE YOUR CRYPTO SECURELY

Once you've purchased or traded some crypto and secured your account, your next priority should be placing it somewhere safe. If you're not leaving it on the exchange to trade later, the only other option is a wallet. Wallets differ in the ownership of your private keys and their connection to the Internet. The choice between them depends on the level of security you're comfortable having.

WHAT IS A PRIVATE KEY?

A private key, like a real key, unlocks your cryptocurrency for you to spend. **Keeping your private key and access to it safe is the most important part of your overall security.** The key is just a really long number – so large that it would be impossible for anyone to guess. If you flip a coin 256 times and write down "1" for heads, "0" for tails, you'll end up with a private key. Here's one we've just generated. It's encoded in hexadecimal (using numbers 0-9 and characters a-f) for a more compact representation:

8b9929a7636a0bff73f2a19b1196327d2b7e151656ab2f515a4e1849f8a8f9ba

If you look that number up on Google, you'll see the only occurrence is in this article (unless it's been subsequently copied elsewhere). That should give you an idea of how truly random the number is – the odds of anyone having ever seen it before are astronomically low.

That example still doesn't do it justice. The number of possible private keys is close to the number of atoms in the known universe. In a nutshell, this is a vital security principle in cryptocurrencies like Bitcoin and Ethereum. Your coins are safe because they're hidden in a brain-meltingly large range.

If you've received funds before, you'll be familiar with public addresses, which are also strings of random-looking numbers. Those are obtained by doing some cryptographic magic on your private key to get a public key, which is hashed to get the public address.

We won't get into depth on how this is done in this article. All you need to know is that, while it's easy to generate a public address with the private key, doing the reverse is impossible today. That's why you can safely list your public address on blogs, social media, etc. No one can spend the funds sent to it without the corresponding private key.

If you lose your private key, you lose access to your funds. If someone else learns your key, they can spend those funds. **As a result, keeping your private key away from prying eyes is of paramount importance.**

SEED PHRASES

You should note that wallets today rarely have just one private key – they're hierarchical deterministic (HD) wallets, meaning they can hold billions of different keys. All you need to know is a seed phrase, a collection of human-readable words that can be used to generate those keys. It may resemble the following:

Strike sadness boss daring voice connect holiday vintage quantum pony stable genuine

Unless you deliberately choose to use only one private key, you'll probably be asked to back up a seed phrase when you create a new wallet. When we discuss key storage later, the term keys will be used interchangeably to describe both private keys and seeds.

HOW TO SECURE YOUR SEED PHRASE

Your 12, 18, or 24-word seed phrase is extremely important to keep secure and safe. Anyone who has access to the phrase can import your keys into their wallet and steal your funds. You may also have a JSON file or individual private keys that act the same as a seed phrase. Think extremely carefully about how you manage your keys by following our tips below.

1. **Keeping your seed phrase saved on a device connected to the Internet isn't recommended.** If you download a virus or your computer is hacked and controlled remotely, your phrase can be compromised.

2. **Offline storage is much more secure.** You could store the phrase physically or on an offline device. Even if you have a cold storage device that we'll discuss later, you should also backup the key if your device breaks.

3. **If you decide to store your phrase physically, think about the material you'll use and where you'll keep it.** Writing the words on a piece of paper that can be destroyed or easily lost at home isn't a good idea. You might want to use a safety deposit box in a secure location or store the phrase with your bank. Some people will even engrave their seed phrase onto metal as it can't be easily destroyed or use metal letters on a seed board.

For example, giving PancakeSwap permission allows it to automate transactions like adding multiple tokens to a liquidity pool. The DApp can complete different steps all in one go, saving you time. While this is useful, there are some risks associated with it.

Unless you've studied the smart contract yourself and understand exactly what it does, there's always a chance of a backdoor exploit. Typically, projects go through auditing to prove that their smart contracts are safe. Certik is a famous provider of audits, but this reputation still doesn't always guarantee safety.

A compromised project will ask for permission to move unlimited or large amounts of tokens. Less experienced users are more likely to accept these and become victims of fraud. Even if you remove your funds from the DeFi platform, the project may still have some control and be able to steal them. Hackers can also attempt to manipulate and abuse smart contracts. Once again, if you've given permission to a project, you could be at risk in this situation.

USE AUDITED PROJECTS THAT OFFER MORE SECURITY

As we mentioned above, audited projects are more secure options to invest your tokens and coins with. If you're interacting with smart contracts, staking in pools, or providing liquidity, it's recommended you always look for projects with audits.

An audit analyses a DApp's smart contract code. The auditors will look for backdoors, exploitable scripts, and security issues. These are reported to the project founders, who then make changes to the code. Any changes are added to the final report to show users the complete, transparent process. The final report can then be made public.

While an audit cannot guarantee a project's safety, the chance of your funds being more secure does improve. It would be unwise to invest money in a project that has no audit available. Some smart contracts handle a massive amount of funds which makes them attractive to hackers. If auditors don't check the code, they become easy targets.

Certik regularly updates their list of audited projects, along with their rating out of 100 and other important information.

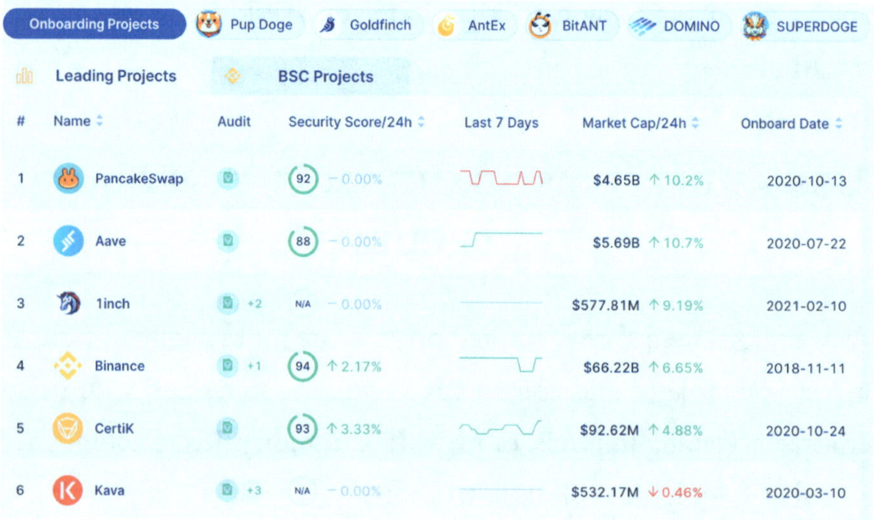

#	Name	Audit	Security Score/24h	Last 7 Days	Market Cap/24h	Onboard Date
1	PancakeSwap		92 – 0.00%		$4.65B ↑ 10.2%	2020-10-13
2	Aave		88 – 0.00%		$5.69B ↑ 10.7%	2020-07-22
3	1inch	+2	N/A – 0.00%		$577.81M ↑ 9.19%	2021-02-10
4	Binance	+1	94 ↑ 2.17%		$66.22B ↑ 6.65%	2018-11-11
5	CertiK		93 ↑ 3.33%		$92.62M ↑ 4.88%	2020-10-24
6	Kava	+3	N/A – 0.00%		$532.17M ↓ 0.46%	2020-03-10

How to avoid scams

Cryptocurrencies, unfortunately, attract many scammers. People look to exploit other users and take their crypto, and once the funds are stolen, there is usually no way of getting them back. Scammers abuse the anonymous nature of cryptocurrencies and the fact that many users directly control large amounts of funds.

You should always be vigilant and never send money to users you don't know. You should also always check the identity carefully of anyone you do send money to. Here are some of the most common scams to look out for:

1. **Phishing** - You may receive an email from an exchange or other service you use, asking you to log in or provide personal information. However, this may be a scammer looking to steal your information.

2. **Fake exchanges** - These are often mobile apps or websites which imitate the look of an exchange. Once you enter your details, a scammer will then use it to access your real account.

3. **Blackmail** - A scammer may send you malware that holds your files for ransom. To pay, you will most likely have to send Bitcoin or another currency to get them back. You may not even receive the files after payment.

4. **Pyramid and Ponzi schemes** - You may be offered to participate in a new project and purchase its coins or enter a

special deal requiring you to make a crypto payment. However, a deal that's too good to be true often is. Do your own research to make sure what you're investing in is safe.

5. **Impersonation** - Someone may pretend to be an official, person of trust, or even friend. They will then ask you for crypto or information that you would not typically give out. In this case, always double-check someone is who they say they are.

When it comes to keeping your cryptocurrencies secure, the blockchain industry today provides many security measures. From trading through to storing and using your crypto, simple tips are effective in keeping your funds safe. In terms of storage, each alternative has its benefits and drawbacks, so it's essential to understand the trade-offs. As always, make sure to do proper research into anywhere you're putting your money or crypto.

Ledger Nano X specifications

Dimensions (cm)	7.2 x 1.86 x 1.18
Weight	34 g
Screen	128 x 64 OLED
Input	Two buttons
Connection	USB-C, Bluetooth
Battery	100 mAh Lithium-ion
Compatibility	• 64-bit PCs (Windows 8+, macOS 10.8+, Linux), excluding ARM processors • iOS 9+ or Android 7+ smartphones
Supported coins and tokens	~1200+

What's in the box?

- Ledger Nano X
- USB Type-C cable
- Getting started leaflet

- 3 recovery sheets
- Keychain strap

LEDGER NANO X OVERVIEW

The Ledger Nano X is a compact device that easily fits into a pocket, making it convenient for you to manage your crypto holdings on the go. Compared to the Ledger Nano S, it has a slightly larger screen and better-placed buttons that are easier to press.

You can set up the Ledger Nano X using a desktop computer, or an iOS or Android device. Throughout our testing, the initial setup process was quite smooth and intuitive – suitable for even absolute beginners. Pairing the device using both USB-C and Bluetooth worked consistently and without any complications. According to Ledger, the battery life is about 8 hours, and the battery should have about a 5-year lifespan.

The most convenient way to manage crypto holdings on a Ledger device is through the Ledger Live app – available for both desktop and mobile. It provides all the standard functionality you would expect, such as checking balances, or sending and receiving assets. But, it also offers more advanced features, such as SegWit transactions, exchange integrations, and the ability to export the extended public key (xpub).

One of the limitations of the Ledger Nano S is that you can only install a few apps at the same time. In contrast, the Ledger

Nano X can store up to 100 apps at the same time, making it suitable to manage a diversified portfolio.

Ledger Nano X pros and cons

Pros

- It can be set up and used with a smartphone or tablet.
- Very portable and has a built-in battery.
- Supports all major blockchains and a wide range of tokens.
- Easy to use.
- Adding coins is fast and intuitive.
- Supports multiple languages.
- Ledger is a well-established company that has operated since 2014.

Cons

- The Bluetooth function might pose a security risk. But, according to the company, no critical data (such as private keys or seed) are transmitted through Bluetooth.

LEDGER NANO X PRICING

As of November 2021, the Ledger Nano X is sold for $119.

It is advised to always buy hardware wallets directly from the manufacturer. Buying second-hand hardware wallets poses the risk of the private keys being compromised.

The Ledger Nano X is the latest product from one of the most well-known hardware wallet manufacturers in the crypto industry. Its predecessor, the Ledger Nano S has been one of the most popular cryptocurrency hardware wallets due to its ease of use and smooth user experience. The Ledger Nano X comes with an increased capacity for coins and Bluetooth connectivity.

In summary, the Ledger Nano X is an easy-to-use and accessible hardware wallet, suitable for both beginners and advanced users.

Device fingerprints may come in handy when detecting and preventing identity theft or credit card fraud. However, these techniques threaten user privacy and depending on the implementation, the collection of data can be undetectable - especially with passive fingerprinting.

WHAT ARE THE LIMITATIONS?

When it comes to active fingerprinting, the data collection relies on the availability of scripting languages, such as JavaScript. Mobile devices and users that run privacy software or plugins will likely present limited scripting availability, making them harder to identify. This includes the use of browser extensions that block trackers and ads.

In some situations, however, privacy-focused users can be easier to identify. For instance, when they use unpopular software and plugins along with particular settings that, ironically, make them even more distinctive.

Also, the effectiveness of fingerprinting may be limited by large variations on the client-side. Users that are constantly changing their settings, or that use multiple virtual operating systems may cause inaccuracies in the process of data collection.

Using different browsers may as well cause inconsistencies in the process of gathering information, but modern, cross-browser fingerprinting techniques can be used to avoid such limitations.

There are several ways of implementing and using the device fingerprinting technique. So, the effectiveness to gather data and identify a single source may vary considerably from one method to another.

Either on its own or when combined with other methods, device fingerprinting can be an effective tool for tracking behavior and identifying users. As such, this powerful technique can be used for both legit and shady activities, so being informed about their basic mechanisms is certainly a good starting point.

About the Author

Kalei Poteat was born in Honolulu, Hawaii. He is a Communications Major who wants to inspire young adults into new possibilities of growth and development. Committed to greatness, he has started multiple business in several different industries such as logistics, hemp, publishing, and marketing.

Kalei believes that there is a lot someone can do in life when you discover your purpose. He is a reminder that spreading love and positivity can increase the chances of anyone looking to enjoy their journey on planet Earth.

"I love to help develop small businesses and entrepreneurs by helping them realize the power of Digital Products and Assets that can potentially 10X their lives. Join me to discuss ways to help your brand reach 6 figures."

Website – www.KaleiPoteat.com

Twitter – www.Twitter.com/KaleiPoteat

Instagram- www.Instagram.com/ImperialPhysics

Facebook – www.Facebook.com/Aloha.Merch

Made in the USA
Las Vegas, NV
03 November 2026

57424179R00136